PRAISE [partially obscured by sticker]

Animal Liberation Now [partially obscured by sticker]

"I became a vegetarian when I read *Animal Liberation* back in the 1970s. Then and there I stopped eating meat. If I'd read this revised *Animal Liberation Now*, I'd have become a vegan much sooner."

—Jane Goodall, author of *In the Shadow of Man* and founder of the Jane Goodall Institute

"Widely recognized as the foundational text within the animal liberation movement, Peter Singer's *Animal Liberation* opened my eyes to the radical philosophy that all animals are worthy of equal consideration. Singer's latest, *Animal Liberation Now*, will motivate a new generation of readers who are resolutely committed to creating a just society for all."

—Joaquin Phoenix

"In its first incarnation, *Animal Liberation* became the indispensable foundational text for the movement whose name it bore. In its new, updated, and wholly rewritten form, *Animal Liberation Now* provides not only a survey—sober, authoritative, and chilling—of what goes on today in the factory farms and research laboratories of the world, but also a guide, written with the honesty and philosophical depth characteristic of all of Peter Singer's work, through the complexities of the modern debate on animal rights."

—J. M. Coetzee, author *Disgrace* and *The Lives of Animals*

"Peter Singer may be the most moral person on the planet. If his ruthlessly consistent altruism makes the rest of us shuffle our feet in discomfort, or even noisily disrupt his lectures, that's all the more reason to read this book."

—Richard Dawkins, Fellow of the Royal Society, author of *The Selfish Gene*, *The God Delusion*, and *The Ancestor's Tale*

"*Animal Liberation Now* is a truly revolutionary text to read, treasure, and pass along. Peter Singer's *Animal Liberation* was a philosophical bombshell when it first appeared in 1975. It forever changed the conversation about our treatment of animals, and this new book will both enrich and sharpen that conversation."

—Ingrid Newkirk, founder and president of People for the Ethical Treatment of Animals

"*Animal Liberation* has been one of the few great classics in the field of practical ethics. This updated, expanded, and retitled edition, which reflects the growth of Singer's thought over almost half a century, is a landmark in the history of moral philosophy."

—Jeff McMahan, Sekyra and White Professor of Moral Philosophy, University of Oxford

"*Animal Liberation Now* shows that the book that changed the world a half century ago is even more relevant today."

—Dale Jamieson, director, Center for Environmental and Animal Protection, New York University

"Peter Singer has helped transform our sense of what we owe to others, by showing us that the others encompass all sentient beings. As a philosopher, I rejoice at the appearance of *Animal Liberation Now* and thank Singer for bringing honor upon our subject through his life and work."

—Mark Johnston, Henry Putnam University Professor of Philosophy, Princeton University

ANIMAL LIBERATION NOW

ALSO BY PETER SINGER

Why Vegan?

Utilitarianism: A Very Short Introduction (with Katarzyna de Lazari-Radek)

Ethics in the Real World

Famine, Affluence, and Morality

The Most Good You Can Do

The Point of View of the Universe (with Katarzyna de Lazari-Radek)

The Life You Can Save

The Ethics of What We Eat (with Jim Mason)

The President of Good and Evil

Pushing Time Away

One World

Writings on an Ethical Life

A Darwinian Left

Ethics into Action

The Greens (with Bob Brown)

Rethinking Life and Death

How Are We to Live?

Should the Baby Live? (with Helga Kuhse)

Making Babies (with Deane Wells)

Hegel

The Expanding Circle

Animal Factories (with Jim Mason)

Marx

Practical Ethics

Animal Liberation

Democracy and Disobedience

ANIMAL LIBERATION NOW

The Definitive Classic Renewed

With a Foreword by Yuval Noah Harari

PETER SINGER

HARPER PERENNIAL

NEW YORK • LONDON • TORONTO • SYDNEY • NEW DELHI • AUCKLAND

To Richard and Mary, and Ros and Stan,
for enlightening me about our treatment of animals;
To Renata for sharing our decision to stop eating animals,
and for our wonderful journey together;
And to the many, many good people striving to bring about a better world for all sentient beings.

HARPER PERENNIAL

Portions of this book were previously published, in somewhat different form, in earlier editions of *Animal Liberation*.

Designed by Jamie Lynn Kerner

Library of Congress Cataloging-in-Publication Data has been applied for.

ISBN 978-0-06-322670-8 (pbk.)
ISBN 978-0-06-333598-1 (library ed.)

23 24 25 26 27 LBC 7 6 5 4 3

CONTENTS

INTRODUCTION

Yuval Noah Harari

ANIMALS ARE THE MAIN VICTIMS OF HISTORY, AND THE TREATment of domesticated animals in industrial farms is perhaps the worst crime in history. These statements would have sounded ludicrous back in 1975, when Peter Singer first published *Animal Liberation*. Today, thanks in no small part to the impact of this seminal book, increasing numbers of people accept these ideas as reasonable, or at least debatable.

In the decades since the publication of *Animal Liberation*, scientists have increasingly turned their attention to the study of animal cognition, animal behavior, and human-animal relations. What they have discovered has largely confirmed Singer's main insights: The human march of progress is strewn with dead animals. Already tens of thousands of years ago, our Stone Age ancestors were responsible for a series of ecological disasters. When the first humans reached Australia about 45,000 years ago, they quickly drove to extinction 90 percent of its large animals. This was the first significant impact *Homo sapiens* had on the ecosystem. It was not the last.

About 15,000 years ago humans colonized America, wiping out in the process about 75 percent of its large mammals. Numerous other species disappeared from Africa, from Eurasia, and from the myriad islands around their coasts. The archaeological record of country after country tells the same sad story. The tragedy opens with a scene showing a rich and varied population of large animals, without any trace of *Homo sapiens*. In scene two, Sapiens appear, evidenced by a fossilized bone, a spear point, or perhaps a campfire. Scene three quickly follows, in which men and women occupy center stage and most large animals, along with many smaller ones, are gone. Altogether, Sapiens drove to extinction about 50 percent of all the large terrestrial mammals of the planet *before* they planted the first wheatfield, shaped the first metal tool, wrote the first text, or struck the first coin.

The next major landmark in human-animal relations was the Agricultural Revolution. It involved the appearance of a completely new life-form on earth: domesticated animals. Initially it seemed to be of minor importance, as humans managed to domesticate less than twenty species of mammals and birds, compared to countless thousands of species that remained "wild." Yet with the passing of the centuries, this novel life-form became dominant. Today more than 90 percent of all large animals are domesticated. Consider the chicken, for example. Ten thousand years ago it was a rare bird confined to small niches in South Asia. Today billions of chickens live on almost every continent and island, bar Antarctica. The domesticated chicken is probably the most widespread bird in the annals of planet earth. If you measure success in terms of numbers, chickens, cows, and pigs are the most successful animals ever.

Alas, domesticated species paid for their unparalleled collective success with unprecedented individual suffering. The animal kingdom has known many types of pain and misery for millions of years. Yet the Agricultural Revolution created completely new kinds of suffering, which only became worse with the passing of the generations.

At first sight, domesticated animals may seem much better off than their wild cousins and ancestors. Wild buffalos spend their days searching for food, water, and shelter, and are constantly threatened by lions, parasites, floods, and droughts. Domesticated cattle, in contrast, enjoy human care and protection. People provide cows and calves with food, water, and shelter; treat their diseases; and protect them against predators and natural disasters. True, most cows and calves sooner or later find themselves in the slaughterhouse. Yet does that make their fate any worse than the fate of wild buffalo? Is it any better to be devoured by a lion than slaughtered by a man? Are crocodile teeth kinder than human steel blades?

What makes the existence of domesticated farm animals particularly cruel is not just the way they die but, above all, their living conditions. Two competing factors have shaped the lives of farm animals: On the one hand, humans want meat, milk, eggs, leather, animal muscle power, and amusement. On the other hand, humans have to ensure the long-term survival and reproduction of their farm animals. Theoretically, this should have protected animals from extreme cruelty. If a farmer milked his cow without providing her with food and water, milk production would dwindle, and the cow herself would quickly die.

Unfortunately, humans can cause tremendous suffering to farm animals without endangering their survival and reproduction. The root of the problem is that domesticated animals have inherited from their wild ancestors many physical, emotional, and social needs that are redundant in human farms. Farmers routinely ignore these needs, without paying any economic price. They lock animals in tiny cages, mutilate their horns and tails, separate mothers from offspring, and selectively breed monstrosities. The animals suffer greatly, yet they live on and multiply.

Doesn't that contradict the most basic principles of Darwinian evolution? The theory of evolution maintains that all instincts, drives, and emotions have evolved in the interest of survival and reproduction. If so, doesn't the continuous reproduction of farm animals prove that all their real needs are met? How can a cow have a "need" that is not really necessary for survival and reproduction?

It is certainly true that all instincts, drives, and emotions evolved under evolutionary pressure to survive and reproduce. However, when these pressures disappear, the instincts, drives, and emotions they had shaped do not evaporate instantly. Even if they are no longer instrumental for survival and reproduction, they continue to shape the subjective experiences of the animal. The physical, emotional, and social needs of present-day cows, dogs, and humans don't reflect their current conditions but, rather, the evolutionary pressures their ancestors encountered tens of thousands of years ago. Why do modern people love sweets so much? Not because in the early twenty-first century we must gorge on ice cream and chocolate in order to survive. Rather, it is because if our Stone Age ancestors came across sweet ripened fruits, the most sensible thing to do was to eat as many of them as quickly as possible. Why do young men drive recklessly, get involved in violent quarrels, and try their power in hacking confidential internet sites? Not because they wish to defy present-day state law (which forbids all these actions), but because they must obey ancient genetic decrees: 70,000 years ago, a young hunter who risked his life chasing a mammoth outshone all his competitors and won the hand of the local beauty—and we are now stuck with his macho genes.

Exactly the same kind of logic shapes the life of cows and calves in our industrial farms. Ancient wild cattle were social animals. In order to survive and reproduce, they needed to communicate, cooperate, and compete effectively. Like all social mammals, wild cattle learned the necessary social skills through play. Puppies, kittens, calves, and children all love to play because evolution implanted them with this urge. In the wild, they *needed* to play. Otherwise, they could not learn the social skills vital for survival and reproduction. If a kitten or calf were born with some rare mutation that made them indifferent to play, they were unlikely to survive or reproduce. Similarly, evolution implanted puppies, kittens, calves, and children with an overwhelming desire to bond with their mothers. A chance mutation weakening the mother-infant bond was a death sentence.

What happens when human farmers now take a young calf, separate her from her mother, put her in a tiny cage, vaccinate her against various diseases, provide her with food and water, and then, when she is old enough, artificially inseminate her with bull sperm? From an objective perspective, this calf no longer needs either maternal bonding or playmates in order to survive and reproduce. All her needs are being cared for by her human masters. But from a subjective perspective, the calf still feels a very strong urge to bond with her mother and to play with other calves. If these urges are not fulfilled, the calf suffers greatly.

This is the basic lesson of evolutionary psychology: A need shaped thousands of generations ago continues to be felt subjectively even if it is no longer really necessary for survival and reproduction in the present. Tragically, the Agricultural Revolution gave humans the power to ensure the survival and reproduction of domesticated animals while ignoring their subjective needs. In consequence, domesticated animals are collectively the most successful animals in the world, and at the same time they are individually the most miserable animals that ever existed.

The situation only became worse in the last few centuries, when traditional agriculture gave way to industrial farming. In traditional societies such as ancient Egypt, the Roman empire, or medieval China, humans had a very partial understanding of biochemistry, genetics, zoology, and epidemiology. Consequently, their manipulative powers were limited. In medieval villages chickens ran free among the houses, pecked seeds and worms from the garbage heap, and built nests in the barn. If an ambitious peasant tried to lock a thousand chickens inside a crowded coop, a deadly bird-flu epidemic would probably have resulted, wiping out all the chickens, as well as many of the villagers. No priest, shaman, or witch doctor could have prevented it.

Once modern science deciphered the secrets of birds, viruses, and antibiotics, humans could begin subjecting animals to extreme living conditions. With the help of vaccinations, medications, hormones, pesticides, central air-conditioning systems, automatic feeders, and lots of other novel gadgets, it is now possible to cram tens of thousands of chickens into tiny coops and produce meat and eggs with unprecedented efficiency.

The fate of animals in such industrial installations has become one of the most pressing ethical issues of our time, certainly in terms of the numbers involved. For nowadays most of our planet's big animals live in industrial farms. We imagine that earth is populated by lions, elephants, whales, and penguins. That may be true of the National Geographic channel, Disney movies, and children's fairy tales, but it is no longer true of the real world outside the TV screen. The world contains 40,000 lions and 1 billion domesticated pigs; 500,000 elephants and 1.5 billion domesticated cows; 50 million penguins and 20 billion chickens.

In 2009 there were 1.6 billion birds in Europe, counting all wild species together. That same year, the European meat and egg industry raised 1.9 billion chickens. Altogether, the domesticated animals of the world weigh about 700 million tons, compared with 300 million tons of humans, and less than 100 million tons of large wild animals ("large" meaning animals weighing at least a few kilograms).

Hence the fate of farm animals is not an ethical side issue. It concerns the majority of the earth's large creatures: tens of billions of sentient beings, each with a complex world of sensations and emotions, who live and die as cogs in an industrial production line. If Peter Singer is right, industrial farming is responsible for more pain and misery than all the wars of history put together.

The scientific study of animals has so far played a dismal role in this tragedy. The scientific community has used its growing knowledge of animals mainly to manipulate their lives more efficiently, in the service of human industry. Yet this very same knowledge has demonstrated beyond reasonable doubt that farm animals are sentient beings, with intricate social relations and sophisticated psychological patterns. They may not be as intelligent as us, but they certainly know pain, fear, loneliness, and love. They too can suffer, and they too can be happy.

It is high time we take these scientific findings to heart, because as human power continues to grow, our ability to harm or benefit other animals grows with it. For 4 billion years, life on earth was dominated by natural selection. Now it is increasingly dominated by human intelligent design. Biotechnology, nanotechnology, and artificial intelligence will soon enable humans to reshape living beings in radical new ways, which will redefine the very meaning of life. When we come to design this brave new world, we should take into account the welfare of all sentient beings, and not just of *Homo sapiens.*

Animal Liberation Now raises ethical questions that every human should take to heart. Not everyone may agree with Singer's thesis. But given the immense power humankind wields over all other animals, it is our ethical responsibility to debate it carefully.

PREFACE TO THE 2023 EDITION

Animal Liberation was first published in 1975, and quickly became known as the bible of the emerging and rapidly growing animal rights movement. Activists who rescued animals from laboratories or stole revealing videotapes of cruel experiments began leaving a copy of the book behind to show the ethical basis of their actions. The book's strength was its combination of ethical argument and accurate accounts of what happens to animals in laboratories and factory farms, often based on publications written by the experimenters themselves or by researchers investigating how to produce meat, eggs, and dairy products most efficiently. After the book appeared and spurred the growth of the modern animal movement, the pressure grew for changes in the conditions described in the chapters on the use of animals in research, and in factory farms.

Although *Animal Liberation* has never gone out of print since it was first published, and the central ethical argument on which the book rests has stood up well to more than forty years of challenges, much else has changed:

- In 1975, there was no animal rights movement, and anti-cruelty organizations were mostly concerned with dogs and cats. Now millions of people support organizations working to reduce the suffering of farmed animals and animals used in research.

- In a few U.S. states, these organizations have used citizen-initiated referendums to outlaw keeping farmed animals in cages or stalls too small to allow them to stretch their limbs, turn around, or walk a single step; but these extreme forms of confinement still dominate the states that produce the most pigs and have the most laying hens.
- In Europe, more comprehensive changes were implemented by national parliaments, and by the European Union as a whole.
- Political parties focused on justice for animals have won seats in European national parliaments, in the parliament of the European Union, and in state parliaments in Australia.
- The basic treaty of the European Union recognizes that animals are not mere items of property, conferring on them the legal status of a sentient being.
- The media no longer ridicules animal rights activists; mostly, it takes them seriously.
- In 1975, no one knew what the word "vegan" meant. Now it is widespread on restaurant menus and supermarket labels.

Yet this progress is not universal. As China became more prosperous, it has greatly expanded its animal production. It is now the world's largest producer of pigs and a huge producer of chickens and ducks as well. The expansion of China's animal production, many aspects of which are completely unconstrained by any national animal welfare laws, is not showing any signs of slowing. As I write this preface, huge skyscraper "farms," twenty-six stories high, with 400,000 square meters, or more than 4 million square feet, on each floor, are being built. When complete, they will be filled with millions of pigs.

The struggle for animal liberation has made progress since 1975—but we are still failing to prevent atrocities on a vast scale. As we shall see in chapter 3, in 2020, in the United States, a million pigs were locked in sheds to which heat and humidity were then added until they died of heatstroke. Awful as these deaths are, they are just one example of the suffering inflicted, every year, on the 83 billion mammals and birds raised and killed for food, the majority kept confined all their lives in crowded sheds, and never able to go outside. For them, and for all the others who will, unless there is a sudden and radical change, suffer and die in the decades to come, I have written *Animal Liberation Now,* renewing the argument of the original book and applying it to the conditions prevailing in the twenty-first century.

We know much more now about the consciousness of animals and their psychological, as well as physical, needs, than we did when *Animal Liberation* was first published. From orangutans to octopuses, we have come to understand the lives of the remarkable other animals with whom we share this planet. Rigorous scientific research has established that the capacity to feel pain is not limited to mammals and birds, but extends to fish and at least some invertebrates, not only octopuses, but also lobsters and crabs. This new knowledge increases the urgency of extending the circle of our concern, for we are raising and killing fish and other aquatic animals in even larger numbers than we do mammals and birds.

At the same time, we know that our greenhouse gas emissions are changing the climate of our planet, bringing unprecedented heatwaves, forest fires, and floods and putting all sentient beings, ourselves included, in peril. The meat and dairy industries contribute to this catastrophic change, on a scale comparable to that of the entire transport sector, thus providing another powerful reason for the change in our diet that animal rights advocates have long been urging. Ending factory farming would bring other environmental benefits too, cleaning up our polluted rivers, improving the air many rural people breathe, and dramatically reducing deaths from heart disease and cancers of our digestive system.

Animal Liberation Now shows that despite the greater prominence of the animal movement, we continue to mistreat animals on an incomprehensible scale. This book advocates a new ethic for our relations with animals that starts from the premise that they are sentient beings, with lives of their own to live, who have done nothing to deserve the suffering we inflict on them. This book is a call to join with others to bring about a radical change in the way we treat them.

Peter Singer

CHAPTER 1

All Animals Are Equal . . .

or why the ethical principle on which human equality rests requires us to extend equal consideration to animals too

THE BASIS OF EQUALITY

"ANIMAL LIBERATION" MAY SOUND MORE LIKE A SPOOF OF OTHER liberation movements than a serious objective. In fact, the idea of "The Rights of Animals" was once used to parody the case for women's rights. When Mary Wollstonecraft, a forerunner of modern feminists, published her *Vindication of the Rights of Woman* in 1792, her views were widely regarded as absurd, and before long an anonymous publication appeared titled *A Vindication of the Rights of Brutes*. The author of this satirical work (now known to have been Thomas Taylor, a distinguished Cambridge philosopher) tried to refute Mary Wollstonecraft's arguments by showing that they could be carried one step further. If the argument for equality was sound when applied to women, why should it not be applied to dogs, cats, and horses? Yet to hold that these "brutes" had rights was manifestly absurd. Therefore the reasoning for the equality of women must also be unsound.

Let us assume that we wish to defend the case for women's rights against Taylor's attack. How should we reply? One way would be by saying that the case for equality between men and women cannot validly be extended to nonhuman animals. Women have a right to vote, for instance, because they are just as capable of making rational decisions about the future as men are; dogs, on the other hand, are incapable of understanding the significance of voting, so they should not have the right to vote. There are many other capacities that men and women share, while humans and animals do not. So, it might be said, men and women are equal and should have equal rights, while humans and nonhumans are different and should not have equal rights.

This reasoning is correct as far as the case for equality between men and women is concerned. The important differences between humans and other animals must give rise to some differences in the rights that each have. But there are also important differences between adults and children. Since neither dogs nor young children can vote, neither has the right to vote. Recognizing this, however, does not count against extending a more basic principle of equality to children, or to nonhuman animals. That extension does not imply that we must treat everyone in exactly the same way, regardless of age or mental capacity. The basic principle of equality does not require equal or identical treatment; it requires equal consideration. Equal consideration for different beings may lead to different treatment and different rights.

So there is a different way of replying to Taylor's attempt to parody the case for women's rights that does not deny the obvious differences between human beings and nonhuman animals but finds nothing absurd in the idea that the basic principle of equality applies to so-called brutes. At this point such a conclusion may appear unwarranted, but if we examine more deeply the basis on which our support for the equality of all humans rests, we will see that we would be on shaky ground if we were to demand equality for every member of the species *Homo sapiens* while denying equal consideration to nonhuman animals.

To make this clear we need to see, first, exactly what it is that we are asserting. Those who wish to defend hierarchical, inegalitarian societies have often pointed out that by whatever test we choose it simply is not true that all humans are equal, in the descriptive sense of that word. Humans come in different shapes and sizes; they come with different intellectual abilities, different physical strengths, different moral capacities, different degrees of sensitivity to and compassion for the needs of others, different abilities to communicate effectively, and different capacities to experience pleasure and pain. In short, if the demand for equality were based on the actual equality of all human beings, we would have to stop demanding equality.

Fortunately, there is no logically compelling reason for assuming that a factual difference in ability between two people justifies any difference in the moral weight we should give to their needs and interests. Equality is a moral ideal, not an assertion of fact. The principle of the equality of human beings is not a description of an alleged actual equality among humans: it is a prescription of how we should treat human beings.

Jeremy Bentham, the founder of the reforming utilitarian school of moral philosophy, incorporated the essential basis of moral equality into his system of ethics by means of the formula: "Everybody to count for one, nobody for more than one." In other words, the interests of every being affected by an action are to be considered and given the same weight as the like interests of any other being. John Stuart Mill said that the first principle of utilitarianism is "perfect impartiality between persons." A later utilitarian, Henry Sidgwick, put the point in this way: "The good of any one individual is of no more importance, from the point of view (if I may say so) of the Universe, than the good of any other." R. M. Hare, who held the chair of moral philosophy at the University of Oxford when I was a student there in the 1970s, argued that to make an ethical judgment sincerely, one must be willing to put oneself in the position of all of those affected by one's decision, and still wish that the judgment be carried out. John Rawls, a professor at Harvard University and the most prominent American philosopher working in ethics in the same period, captured a similar idea with the device of a "veil of ignorance," behind which people must choose the principles of justice to govern the society in which they will live. Only after the principles have been decided upon, and the veil is lifted, will they find out what characteristics they have and what positions they occupy.[1]

It is an implication of this principle of equality that our concern for others and our readiness to consider their interests ought not to depend on what they are like or on what abilities they may possess. Precisely what our concern requires us to do may vary according to the characteristics of those affected by what we do: Concern for the well-being of children requires that we teach them to read; concern for the well-being of pigs may require no more than that we leave them with other pigs in a place where there is adequate food and room to roam freely. The basic element is taking into consideration the interests of the being, whatever those interests may be, and this consideration must, according to the principle of equality, be extended equally to all beings with interests irrespective of their race, sex, or species.

It is on this basis that the cases against racism and sexism must ultimately rest; and it is in accordance with this principle that speciesism must also be condemned. Speciesism, in its primary and most important form, is a prejudice or bias in favor of the interests of members of one's own species and against those of members of other species, on the basis of species alone. A secondary form of speciesism occurs when we give more weight to the interests of some nonhuman animals of a particular species—dogs, for example—than we give to animals with similar interests but of a different species, such as pigs.[2]

BENTHAM'S QUESTION

MANY THINKERS HAVE PROPOSED THE PRINCIPLE OF EQUAL CONSIDERATION of interests, in some form or other, as a basic moral principle; but few have recognized that this principle applies to members of other species as well as to our own. Jeremy Bentham was one of the exceptions. In a forward-looking passage, composed at a time when slaves of African descent had been freed by the French but were still enslaved in the British dominions, Bentham wrote:

> The day may come when the rest of the animal creation may acquire those rights which never could have been withholden from them but by the hand of tyranny. The French have already discovered that the blackness of the skin is no reason why a human being should be abandoned without redress to the caprice of a tormentor. It may one day come to be recognized that the number of the legs, the villosity [having fur] of the skin, or the termination of the *os sacrum* [having a tail] are reasons equally insufficient for abandoning a sensitive being to the same fate. What else is it that should trace the insuperable line? Is it the faculty of reason, or perhaps the faculty of discourse? But a full-grown horse or dog

> is beyond comparison a more rational, as well as a more conversable animal, than an infant of a day or a week or even a month, old. But suppose they were otherwise, what would it avail? The question is not, Can they *reason*? nor Can they *talk*? but, Can they *suffer*?[3]

In this passage Bentham points to the capacity for suffering as the vital characteristic that gives a being the right to equal consideration. The capacity for suffering—or, more strictly, for suffering and/or pleasure or happiness—is not just another characteristic like the possession of reason or language, self-awareness, or a sense of justice. Those who draw "the insuperable line" with reference to these characteristics are including some beings with a capacity for suffering and excluding others. Bentham, on the other hand, is saying that we must consider the interests of *all* beings with the capacity for suffering or enjoyment. He is not excluding any interests from consideration, because the capacity for suffering and enjoyment is a prerequisite for having interests, a condition that must be satisfied before we can properly speak of interests at all. It would be absurd to say that it was not in the interests of a stone to be kicked along the road by a child. A stone does not have interests because nothing that we can do to it could possibly make any difference to its welfare. The capacity for suffering and enjoyment is, however, not only necessary but also sufficient for us to say that a being has interests—at a minimum, an interest in not suffering. Mice, for example, do have an interest in not being kicked along the road because they will suffer if they are treated in that way.

Although Bentham speaks of "rights" in the passage I have quoted, the argument is really about equality rather than rights. In a different passage, Bentham famously described "natural rights" as "nonsense" and "natural and imprescriptible rights" as "nonsense upon stilts." When he refers to moral rights, he is advocating protections for people and animals that ought to be recognized by law and by public opinion; but the real weight of the moral argument does not rest on rights, for they in turn have to be justified on the basis of their tendency to reduce suffering and increase happiness, not only in individual cases but in the long run and for all those affected. We can, therefore, argue for equality for animals without getting embroiled in philosophical controversies about the grounds for rights, or who has them, and which rights they have. The language of rights is a convenient political shorthand that is even more valuable now, in the era of the eight-second soundbite, than it was in Bentham's day; but it is not essential to the argument for a radical change in our attitude toward animals.

If a being suffers, there can be no moral justification for refusing to take that suffering into consideration. No matter what the nature of the being, the principle of equality requires that its suffering be counted equally with the like suffering—insofar as rough comparisons can be made—of any other being. If a being is not capable of suffering, or of experiencing pleasure or happiness, there is nothing to be counted. So the limit of sentience (using the term to indicate the capacity to experience pain or pleasure) is the only defensible boundary of concern for the interests of others.

Are we not justified, though, in giving greater weight to the interests of other humans—because they are members of our own species, whereas other animals are not? Such claims—that "*we* are . . . (insert the name of the group we identify with) and *they* are not"—have been used previously as a justification for refusing to give equal consideration to the interests of others. Racists violate the principle of equality by giving greater weight to the interests of members of their own race, and sexists violate it by favoring the interests of their own sex. Today we can recognize such purported justifications for racism and sexism as spurious ideologies that were accepted only because they served the interests of the dominant group. Similarly, speciesists allow the interests of their own species to override the greater interests of members of other species. In each case there is a dominant group that sees those outside it as inferior in order to justify using them as they, the dominant ones, wish.

You might be thinking: No, humans are different! We are smarter than animals, we are rational beings, we are self-aware, we plan ahead, we are free, we are moral agents. Hence we have rights that other animals do not have, and we are entitled to use other animals as we wish. But, as Bentham pointed out, this argument has the implication that human infants—who are less rational, less self-aware, and less capable of planning ahead than many nonhuman animals—would also lack rights, and we would thus be entitled to use them as we use animals. The same would hold for some humans beyond infancy who, whether from a genetic abnormality, or brain damage, will never match the cognitive capacities of some nonhuman animals. Moreover, a series of studies led by Harvard psychology researcher Lucius Caviola, together with other researchers in psychology and philosophy at Oxford and Exeter universities, has shown that differences in mental capacities don't really explain why people give humans moral priority over other animals. On the contrary, when subjects were asked to say whom they would help in a situation in which they must choose between a human and a chimpanzee, 66 percent chose to help the human, even when told that the chimpanzee had more advanced mental capacities than the human.[4]

In the first edition of this book, I discussed the view put forward by Stanley Benn, a widely respected philosopher who spent most of his academic career at the Australian National University, that we should treat beings according to what is "normal for the species" rather than according to their actual characteristics, and that it is therefore justifiable to give priority to humans over animals, even in a situation in which the human has a lower mental capacity than the animal. Since then other philosophers have defended similar views.[5] The research carried out by Caviola's team shows, however, that the typical mental capacity level of a species had no significant effect on the answers that the participants in their study gave to questions about whom they would help. This does not, of course, refute the ethical claim that this should make a difference, but it does show that this doesn't seem to be an important factor in why people think that humans have a higher moral status. I will return to the ethical claim in chapter 6.[6] Caviola and his colleagues conclude that "the central driver of moral anthropocentrism is speciesism."

WHO CAN SUFFER?

MOST HUMAN BEINGS ARE SPECIESISTS. THE FOLLOWING CHAPTERS show that ordinary human beings—not a few exceptionally cruel or heartless humans, but the overwhelming majority—are complicit in the continuation of practices that thwart the most important interests of nonhuman animals in order to promote far less significant human interests. For completeness, however, before I describe those practices, I should address a question that I sometimes still encounter (though much less frequently now than when I first began discussing the question of animal suffering): How do we know that animals other than humans feel pain?

The first step toward answering this question is to ask: How do we know if *anyone*, human or nonhuman, feels pain? We know, from our own experiences, that we ourselves can feel pain.

But how do we know that anyone else feels pain? We cannot directly experience anyone else's pain, whether that "anyone" is our best friend or a stray dog. Pain is a state of consciousness, a "mental event," and as such it can never be observed. Pain in others can only be inferred from behavior like writhing, screaming, or drawing one's hand away from a flame; or perhaps from a brain-imaging device that indicates what is happening in the relevant parts of our brain.

In theory, we could always be mistaken when we assume that other human beings feel pain. It is conceivable that one of our close friends is really a robot, programmed to show all the signs of feeling pain but really no more sensitive than any other intelligent machine. Although this possibility presents a puzzle for philosophers, not one of us has the slightest doubt that our close friends feel pain just as we do. This is an inference, but a perfectly reasonable one, based on observations of their behavior in situations in which we would feel pain, and on the fact that we have every reason to assume that our friends are beings like us, with nervous systems that can be assumed to function as ours do and produce similar feelings in similar circumstances.

If it is justifiable to assume that other human beings feel pain as we do, is there any reason why a similar inference should not be justified in the case of other animals? The Cambridge Declaration on Consciousness, issued by an international group of prominent neuroscientists meeting in Cambridge in 2012, confirms that although human beings have a more developed cerebral cortex than other animals, this part of the brain is concerned with thinking functions rather than with basic impulses, emotions, and feelings. As the Declaration states, "The weight of evidence indicates that humans are not unique in possessing the neurological substrates that generate consciousness. Nonhuman animals, including all mammals and birds, and many other creatures, including octopuses, also possess these neurological substrates."[7]

Nearly all the external signs that lead us to infer pain in other humans can be seen in other animals. The behavioral signs of pain vary with the species but may include writhing, facial contortions, whimpering, moaning, yelping or other forms of calling out, the appearance of fear at the prospect of the pain being repeated, attempts to avoid the source of pain—and to avoid places where pain occurred previously and instead seek places where the animals had only positive experiences—and so on. In addition, we know that other mammals have nervous systems similar to ours, which respond physiologically as ours do when the animal is in circumstances in which we would feel pain: an initial rise of blood pressure, dilated pupils, perspiration, an increased pulse rate, and, if the stimulus continues, a fall in blood pressure. Moreover, when these animals are given analgesics—the same forms of pain relief we take—both their pain behavior and the physiological indicators of pain are reduced. For example, in an experiment led by T. C. Danbury, of the Department of Clinical Veterinary Science at the University of Bristol, chickens taken from commercial flocks were offered two differently colored foods, one of which contained carprofen, an anti-inflammatory drug. Birds who were lame (which, as we shall see in chapter 3, is common among commercially raised chickens) chose more of the food with carprofen, and their limping decreased in proportion to the dose they consumed, thus showing a close parallel with the effect that relieving pain has in humans and indicating that the injured birds were likely to be experiencing pain when walking.[8]

The nervous systems of other animals were not artificially constructed as a robot might be constructed to mimic the pain behavior of humans. They evolved, no doubt, because a capacity to feel pain enhances animals' prospects of survival, leading them to avoid sources of injury and death. Most of this evolution took place among the ancestors we share with other vertebrates, before we diverged from them.

It has long been accepted as sound policy in science to search for the simplest possible explanation of whatever it is we are trying to explain. It is simpler to suppose that nervous systems that are virtually identical physiologically, have a common origin and a common evolutionary function, and result in similar forms of behavior in similar circumstances, also operate in a similar manner—that is, by giving rise to similar conscious experiences—than it is to hold that despite all these scientifically demonstrable parallels, when it comes to subjective feelings, our nervous system operates in a manner that is entirely different from the way the nervous systems of other vertebrates operate.

When I wrote the first edition of this book, psychological research on animals was just starting to emerge from a period dominated by a form of behaviorism based on the belief that science should refer only to what can be observed. It was thought to be "unscientific" to explain the behavior of animals by referring to the animals' conscious feelings, desires, or purposes. To avoid using terms like "pain" that refer to mental states, the behaviorists filled scientific journals with papers saying that the rats or dogs on whom they were using electric shocks would, when shocked, display "aversive behavior."[9] Then in 1976 Donald Griffin, a researcher with a distinguished record of research in animal behavior, published *The Question of Animal Awareness*, which asked why scientists refused to recognize awareness in nonhuman animals. The book was like a pin to the behaviorist balloon. Once Griffin had raised the question, it became obvious that behaviorist explanations of animal behavior—even something as simple as explaining why a well-fed rat will not pass over an electrified floor in order to obtain food, whereas a half-starved rat will—are more complex than explanations that attribute conscious experiences of pain and hunger to the rat. For we know that any explanation of human behavior in analogous situations—such as a hungry person stealing food despite the risk of punishment—that did not refer to one's mental states, both positive and negative, would be incomplete. Today it seems almost laughable to offer an account of why an animal avoids electric shock without reference to the animal finding the experience painful or unpleasant.

One difference between humans and nonhuman animals is that humans, beyond a certain age and without profound cognitive disabilities, are able to use language and so tell us when they are in pain. Animals are, with some limited exceptions, not able to use language, or at least not one that we can understand. It might be claimed, therefore, that the best evidence we can have that another being is in pain is that they tell us that they are, and so with animals we must remain in doubt, as they cannot tell us. Yet, as Jane Goodall pointed out in her pioneering study of chimpanzees, *In the Shadow of Man*, when it comes to the expression of feelings and emotions, language is less important than nonlinguistic modes of communication such as a cheering pat on the back, an exuberant embrace, a clasp of the hands, and so on. The basic signals we use to convey pain, fear, anger, love, joy, surprise, sexual arousal, and many other emotional states are not specific to our own species.[10] The statement "I am in pain" may be one piece of evidence for the conclusion that the speaker is in pain, but it is not the only possible evidence, and since people sometimes tell lies—and a robot can say "It hurts"—not even the best possible evidence.

Even if there were stronger grounds for refusing to attribute pain to those who do not have a language, the consequences of this refusal might lead us to reject that conclusion. Human infants and very young children are unable to use language. Are we to doubt that a year-old child can feel pain? If not, language cannot be crucial. Of course, most parents understand the responses of their children better than they understand the responses of other animals; but this is just a fact about the greater contact we have with our own infants as compared to animals. Those who have animals as companions soon learn to understand their responses as well as we understand those of an infant—and sometimes better, because the minds of mature dogs and cats are closer to our own than are the minds of infants in the first months of life.

DRAWING THE LINE

THERE IS NO LONGER ANY SERIOUS DISPUTE AMONG SCIENTISTS that at least some nonhuman animals can feel pain and experience other conscious states, both positive and negative. The livelier scientific controversies now are about which animals are, or may be, capable of conscious experience. The beings to whom zoologists apply the term "animal" range from mammals to sea sponges, and there is no good reason to believe that the boundary between animals and plants coincides with the line between beings who can suffer and beings that cannot. Hence, in order to know when the principle of equal consideration of interests applies, we need to know where to draw that line.

Scientists studying pain in animals have developed experimental methods that include examining whether the behavior suggestive of pain that follows a painful stimulus is reduced by the administration of drugs similar to those that alleviate pain in humans. They may also look for evidence of an animal engaging in trade-offs between a painful experience, or the risk of one, and an opportunity for a reward, such as food. They see this kind of flexible decision-making as indicative of a centralized processing of information that involves a common measure of value.[11] This kind of evidence is, however, still only available for a few species.

As we have just seen, with mammals and birds, the evidence for a capacity to feel pain is overwhelming. Among other vertebrates, we catch or raise, and kill, vastly more fish than we do reptiles or amphibians. For that reason, I will just note in passing that there is evidence that reptiles and amphibians can suffer[12] and consider in more detail the evidence about fish.

Fish

Victoria Braithwaite, professor of fisheries and biology at Pennsylvania State University until her death in 2019 and the author of *Do Fish Feel Pain?*, was one of the first scientists to investigate the nervous systems of fish, as well as their behavior in situations that might cause them pain. Her team was the first to show that fish have nociceptors, the sensory receptors that have been shown, in mammals and birds, to detect signals from damaged tissue. She also examined the behavior of fish when they experience something that would cause us physical pain: for example, an injection of vinegar or bee venom into the lips. She found that these stimuli cause fish to behave in ways suggestive of pain, including more rapid breathing, rubbing of the lips, and disregarding other things that are going on in their tank to which they would normally react. These changes in behavior may last several hours, but giving the fish painkillers such as morphine hastened a return to more normal behavior. Fish learn to avoid unpleasant experiences, like mild electric shocks, but they may also engage in the motivational trade-offs mentioned previously, and not only for food: Some fish will choose to endure electric shocks in order to be close to a companion.

These observations indicate that fish satisfy the most important criteria for sentience, but it is also worth dispelling some of the myths about fish lacking cognitive capacities that mammals and birds have. Some species show reciprocal cooperation, taking it in turns to be vigilant for predators while their companion searches for food. Some fish engage in cooperative hunting, something that previously only social mammals had been observed to do, for it suggests both forethought and communication. More remarkably still, this cooperation can be between fish of different species. When a grouper, a large fish, is chasing prey who takes refuge in a crevice too narrow for the grouper to enter, groupers have been observed to swim over to a crevice in which a moray eel is likely to be hiding. The grouper then makes distinctive body movements that lead the eel to leave the crevice and swim with the grouper. When they reach the crevice where the prey fish is hiding, the grouper uses its body to point to it. The eel then enters the crevice. Often, this will flush the prey into open water, where the grouper is waiting to pounce. On other occasions the eel will be able to catch and eat the prey. In contrast to the cooperative hunting of social mammals, the carcass of the kill cannot be shared, because both the grouper and the eel swallow the prey whole. But with repeated occasions for cooperation, both the eel and the grouper benefit.

After summarizing the evidence from her own research and that of others, Braithwaite wrote: "Given all of this, I see no logical reason why we should not extend to fish the same welfare considerations that we currently extend to birds and mammals."[13] I agree, with the caveat that there are about 33,000 species of fish, or roughly five times the number of species of mammals, and research relating to the capacity to feel pain has been done on only a few of them. Braithwaite pointed out that scientists examining cartilaginous fish—sharks and rays—have not found the kind of nociceptors that mammals, birds, and fish with bony skeletons have.[14] So, if we are to be cautious, perhaps we should limit the conclusion that fish are sentient beings to bony fish, also known as teleosts. This does not mean, of course, that sharks and rays are *not* sentient beings, but only that the evidence that they are is not as strong as it is for teleosts, who constitute the overwhelming majority of the fish humans eat.

Why, then, do we treat fish as if they cannot feel anything? Is it because they cannot yelp or whimper, and they do not have facial expressions that we can recognize as indicating distress? Otherwise, surely only a psychopath could enjoy an afternoon spent sitting by a river dangling a barbed hook into the water, while next to them the fish they caught earlier are flapping helplessly as they slowly suffocate and die.

Invertebrates

Just as the majority of vertebrates are fish, not mammals, so the majority of animals are not vertebrates. Invertebrates are an extraordinarily diverse group, which isn't really surprising, given that they are defined only by the absence of a vertebral column. That we do define them in that way is another example of our anthropocentric perspective: There are beings with vertebrates, like us, and then there are all the rest. If we regard invertebrates objectively, we will recognize that some are both sentient and intelligent, and there are vast numbers of others for whom the possibility of sentience cannot be excluded.

When we meet an intelligent vertebrate, whether it is a chimpanzee, an elephant, or a dolphin, we are in the presence of a mind that shares a common ancestor with our own. An octopus, however, is a mollusk, and therefore more closely related to an oyster than to any vertebrate. To find the common ancestor we share with an octopus we have to go back 500 million years to a worm that is unlikely to have been conscious at all. Yet octopuses are intelligent. Several YouTube videos show them solving novel problems like opening a screw-top jar to get at the tasty morsel they can see inside it. There are many anecdotes of them escaping from their tanks, and of getting out of their tanks at night and into neighboring tanks containing fish before returning to their own tank—their version of raiding the cookie jar. Free-living octopuses have learned to use empty half coconut shells to hide in, sometimes transporting the shells considerable distances to the places where the octopuses want to use them, which appears to indicate a capacity to plan ahead.[15] So, if you ever meet an octopus, remember that the meeting is not between minds that are kin, but between minds that evolved entirely independently. It is the closest you are likely to come to contact with an intelligent alien.[16]

The other group of invertebrates on which there is now strong evidence of sentience is decapod crustaceans, a group that includes crabs, lobsters, crayfish, and some shrimps. (The term *decapod* comes from the Greek for "ten legs.") An interdisciplinary team of scientists led by Jonathan Birch, of the London School of Economics, researched the sentience of both cephalopods (the group that includes octopuses and squids) and decapod crustaceans, and summed up their findings in *Review of the Evidence of Sentience in Cephalopod Molluscs and Decapod Crustaceans*, a report submitted to the United Kingdom's Department for Environment, Food and Rural Affairs, which at the time was reviewing the scope of that country's Animal Welfare Act. The researchers examined more than 300 scientific studies of sentience in these animals and developed a set of criteria for establishing whether animals are sentient. They had high confidence in the presence of sensory receptors for pain in decapods, and very high confidence that the brains of crabs and lobsters are capable of integrating information from different sources, though evidence was lacking for other decapods. Chemical compounds, in some cases produced by the animals themselves and in other cases applied externally as part of an experiment, reduced responses to painful stimuli in crabs and lobsters, and in some shrimps. Overall, the researchers found the evidence for sentience in decapod crustaceans to be less strong than that for sentience in cephalopods, but they emphasized that this was at least partially due to a lack of evidence because fewer studies have been done, rather than to evidence against sentience in decapods.

The report, which was released in November 2021, had a central recommendation: "That all cephalopod molluscs and decapod crustaceans be regarded as sentient animals for the purposes of UK animal welfare law." Parliament was at the time debating new legislation recognizing animals as sentient beings; the recognition of animals as sentient beings had been incorporated into law in the European Union, but this status ceased to have legal effect in the United Kingdom following that country's exit from the Union. The UK Animal Welfare (Sentience) Act, which became law in 2022, shows the influence of the report by Birch's team. The act states that for the purposes of the legislation, "animal" means any vertebrate (other than *Homo sapiens*), any cephalopod mollusk, and any decapod crustacean. Octopuses, squid, crabs, lobster, and crayfish are also protected by the animal welfare legislation of New Zealand, Norway, and Switzerland.[17]

We know less about sentience in other crustaceans because there are many species on which no research has been done, and this uncertainty applies even more strongly to insects. Some insect behavior is difficult to reconcile with the idea that insects feel pain. When the female praying mantis ceases to think of the male as a lover and instead treats him as dinner, that does not put an end to his interest in sex with her. Other insects continue to walk on legs that have been crushed and to eat while they themselves are being eaten. In an influential article published in 1984, C. H. Eisemann and several colleagues drew from these examples the conclusion that insects are unlikely to feel pain.[18] Thirty-five years later, Shelley Adamo, a Canadian scientist specializing in invertebrate behavior and physiology, reached a similar conclusion, doubting whether insects have sufficient neurons to support consciousness.[19]

Other scientists are more open to the idea of insect consciousness. As early as 1923, the Austrian scientist Karl von Frisch discovered that honeybees communicate the direction to and distance from food sources by a "waggle dance." Although his initial claims were greeted with skepticism, eventually they gained acceptance, culminating with the award of the Nobel Prize to von Frisch in 1973. But do these complex forms of communication indicate consciousness? Andrew Barron, a neuroscientist specializing in the neural mechanisms of natural animal behavior, and Colin Klein, a philosopher interested in consciousness, have teamed up to argue that the structure of insect brains indicates that they do possess "a capacity for subjective experience."[20]

IMPLICATIONS

WE HAVE NOW REACHED TWO IMPORTANT CONCLUSIONS: MANY animals can feel pain, and there is no moral justification for treating their pain as less important than similar amounts of pain felt by humans. What practical consequences follow from this conclusion? To prevent misunderstanding, I shall spell out what I mean a little more fully.

If I give a horse a hard slap across its rump with my open hand, the horse may start, but they presumably feel little pain; their skin is thick enough to protect them against a mere slap. If I slap a baby in the same way, however, the baby will cry and presumably feel pain, for their skin is more sensitive. This makes it worse, other things being equal, to slap a baby than a horse. But there must be some kind of blow—perhaps a blow with a heavy stick—that would cause the horse as much pain as we cause the baby by slapping them with an open hand. That is what I mean by "the same amount of pain," and if we consider it wrong to inflict that much pain on a baby for no good reason, then we must, if we are to avoid speciesism, consider it equally wrong to inflict the same amount of pain on a horse for no good reason.

Other differences between humans and animals cause other complications. Normal adult human beings have mental capacities that will, in certain circumstances, lead them to suffer more than animals would in the same circumstances. If, for instance, we were to perform lethal scientific experiments on normal adult humans, kidnapped at random for this purpose, this would lead to widespread fear. The same experiments performed on nonhuman animals would cause less suffering since animals are not able to communicate with distant animals as we can communicate with people far away, and hence the animals would not have the anticipatory dread of being kidnapped and experimented upon. This does not mean that it would be right to perform the experiment on animals, but only that there is a reason for preferring to use animals rather than normal adult human beings.

This argument for using animals doesn't mention species, but it does draw on some cognitive differences between normal adult humans and other animals. Is it therefore a disguised form of speciesism? To show that the argument is genuinely non-speciesist, those who use it need to accept that the same argument gives us a reason for preferring to experiment on profoundly cognitively disabled human beings rather than other adults, since the profoundly cognitively disabled humans would also have no idea of what was going to happen to them. If we reject that implication of the argument, while still allowing experiments on animals, we would be simply preferring members of our own species.

There are many matters in which the superior mental powers of most adult humans make a difference: anticipation and planning for the future, more detailed memory of the past, greater knowledge of what is happening, both to themselves and to others, and so on. Yet these differences can cut both ways. If, for instance, we are taking prisoners in wartime we can explain to them that although they must submit to capture, search, and detention, they will not otherwise be harmed and will be set free at the conclusion of hostilities. In contrast, wild animals cannot distinguish attempts to overpower and capture them from an attempt to kill; the one causes as much terror as the other, and possibly more.

Admittedly, comparisons of the sufferings of different species are difficult to make, and for this reason when the interests of animals and humans clash, the principle of equality doesn't give precise guidance. But for the principle to make a difference to our present behavior, precision is not essential. Many of the things we do to animals cause so much pain to them and yet are so inessential to us, that it is obvious that we are harming them more than we are benefiting ourselves—and on a vast scale too. As Kenny Torella wrote in *Vox* in 2022, "for just about every animal species besides *Homo sapiens*, today is probably the worst period in time to be alive—especially for the species we've domesticated for food: chickens, pigs, cows, and increasingly, fish."[21] From such a low point, it's not difficult to do better.

WHEN IS KILLING WRONG?

UP TO NOW I HAVE SAID A LOT ABOUT INFLICTING SUFFERING ON animals but nothing about killing them. That omission has been deliberate. The application of the principle of equality to the infliction of suffering is, in theory at least, fairly straightforward. Pain and suffering are in themselves bad and should be prevented or minimized, irrespective of the species of the being who suffers. How bad a pain is depends on how intense it is and how long it lasts, but pains of the same intensity and duration are equally bad, whether felt by humans or animals. Deciding when it is wrong to kill a being is more complicated. I have kept, and shall continue to keep, the question of killing in the background because in the present state of human tyranny over other species the more simple, straightforward principle of equal consideration of pain or pleasure is a sufficient basis for identifying and protesting against the many ways in which we abuse and exploit animals. Still, this chapter would be incomplete if it did not say something about killing animals.

Just as most human beings are speciesists in their readiness to cause pain to animals when they would not cause similar pain to humans, so most human beings are speciesists in their readiness to kill other animals when they would not kill human beings. We need to proceed more cautiously here, however, because people hold widely differing views about when it is permissible to kill humans, as the continuing debates over abortion and end-of-life decisions for terminally ill patients attest. Nor have ethicists been able to agree on exactly what it is that makes it wrong to kill human beings.

Consider first the idea that taking an innocent human life is always wrong, often referred to as the view that human life is sacrosanct, or inviolable, and used as a basis for opposition to abortion and euthanasia.[22] Few defenders of this view oppose the killing of nonhuman animals. The belief that human life, and only human life, is sacrosanct is a form of speciesism.

To see this, consider the case of Baby Theresa, born in Florida in 1992 with anencephaly, a condition in which most of the brain is missing. Anencephalic infants are not brain dead, because the brain stem, which regulates functions like breathing and the beating of the heart, still functions; but they will never become conscious or be able to smile at their mother. Usually no efforts are made to keep these infants alive, and they die within a few hours of birth. But Baby Theresa's mother, Laura Campo, wanted some good to come from the tragic birth of her child and so offered her as an organ donor, to help another child, perhaps a baby born with a fatal heart defect. (Such babies often die because infant hearts are rarely available for transplantation.) Doctors refused to remove Theresa's organs, however, saying that she was not dead. Campo and the baby's father went to court to seek permission for the organs to be removed while Theresa was still alive, as a heart transplant is less likely to be successful if the heart has stopped beating. The judge denied the request, saying that she could not authorize someone to take a human life, "no matter how short or unsatisfactory" the life may be. Legally, removing the baby's heart would have been murder, and in this respect the law reflects the idea that every innocent human being has an inviolable right to life.

Baby Theresa died a few days later, and her organs benefited no one.[23] Yet the same people who say that killing her would have been wrong would have had no objection to taking a heart from a healthy, fully conscious pig, baboon, or probably even a chimpanzee, if that were a means of saving a human life. How can they justify these different judgments? Some people may appeal to religious views, saying that Baby Theresa had an immortal soul or was made in the image of God, while nonhuman animals lack souls and are not made in God's image. This view lacks any reasoned explanation for the belief that every member of *Homo sapiens* has an immortal soul or is made in the image of God, but no member of any other species is. Important as these beliefs may have been historically, they are less widely held today, and in any case, in a pluralistic community with a separation between church and state, the law should not be grounded on religious beliefs. It seems, therefore, that in such a society, the idea that it would have been wrong to remove a heart from Theresa but right to remove hearts from nonhuman animals with greater awareness and greater potential for enjoying their lives rests on the fact that Theresa was a member of the species *Homo sapiens*, whereas pigs, baboons, and chimpanzees are not. We are, once again, facing a belief that is mere speciesism.[24]

This does not mean that to avoid speciesism we must hold that it is as wrong to kill a dog as it is to kill a human being in full possession of their faculties. The only position that is irredeemably speciesist is the one that tries to make the boundary of the right to life run exactly parallel to the boundary of our own species. To avoid speciesism, we must allow that beings who are similar in all relevant respects have a similar right to life; and mere membership in our own species is not a morally relevant distinction on which to base this right. Within these limits we could still hold, for instance, that it is worse to kill an adult human with a capacity for self-awareness and the ability to plan for the future and have meaningful relations with others, than it is to kill a mouse, who presumably does not share all of these characteristics; or we might appeal to the close and long-lasting family and other personal ties that humans have but mice do not have to the same degree; or we might think that the crucial difference lies in the consequences for other humans, who will be put in fear for their own lives. Whatever criteria we choose, however, we will have to admit that they do not run parallel to the boundary of our own species. We may legitimately hold that there are some features that make it worse to kill most human beings than to kill a nonhuman animal, such as those just stated; but by any non-speciesist standard, many animals possess these features to a higher degree than a human like Baby Theresa or some other humans with conditions that have profoundly and permanently impaired their cognitive abilities. Hence, if we base the right to life on these characteristics, we must grant these animals a right to life at least as strong as we grant to those humans.

This argument could be taken as showing that many nonhuman animals have a strong, perhaps even absolute, right to life and that we commit a grave moral offense whenever we kill them, even when they are old and suffering and our intention is to put them out of their misery. Alternatively, one could take the argument as showing that humans who lack the features that are necessary for a right to life may be killed for quite trivial reasons, as we now kill animals. I have written about these issues at length elsewhere.[25] Because the main concern of this book is with the ethics of how we treat animals, and not with the ethics of end-of-life decisions for humans, I shall not attempt to settle this issue here but will limit myself to saying that while both of the positions just described avoid speciesism, neither is satisfactory. What we need is some middle ground that would avoid speciesism but would not make the lives of profoundly cognitively disabled humans as cheap as the lives of pigs and dogs now are, nor make the lives of pigs and dogs so sacrosanct that we think it wrong to put an end to their suffering when they are terminally ill. What we must do is bring nonhuman animals within our sphere of moral concern and cease to treat their lives as expendable for trivial reasons, and at the same time reconsider our policy of preserving human lives at all costs, even when there is no prospect of a meaningful life or of an existence without unbearable suffering.

To see the difference between the issues of inflicting pain and taking life, think about how we would choose within our own species. If we had to choose to save the life of a human being within the normal cognitive range or a profoundly cognitively disabled human being, then if everything else was equal, most of us would choose to save the life of the cognitively normal human being; but if we had to choose between preventing pain in the cognitively normal human being or the cognitively disabled human being—imagine that both have received painful but superficial injuries, and we only have enough painkiller for one of them—it is not nearly so clear how we ought to choose.

The same is true when we consider other species. The evil of pain is, in itself, unaffected by the other characteristics of the being who feels the pain; while the value of life and the wrongness of killing may be affected by these other characteristics. To take the life of a being who has been hoping, planning, and working for some future goal is to deprive that being of the fulfillment of all those efforts; to take the life of a being with a mental capacity below the level needed to grasp that one is a being with a future—much less make plans for the future—cannot involve this particular kind of loss. It is reasonable to hold that to kill someone who is so strongly oriented toward their long-term future is normally a much more serious wrong than killing a being who lives only in the present and has only a short future-time horizon. (This should, incidentally, indicate how to respond to the common tactic of ridiculing animal advocates by saying that we want to grant mosquitoes the same right to life as humans.[26]) I will not, however, attempt to give a general answer to the question of when it is wrong to painlessly kill an animal. The conclusions argued for in this book flow from the principle of minimizing suffering alone. Interestingly enough, this is true even of the conclusion that we ought, in most circumstances, to avoid eating animal products, a conclusion that in the popular mind is thought to be based on the belief that killing animals is wrong.

LOOKING AHEAD

YOU MAY HAVE SOME QUESTIONS:

- What should we do, if we accept the principle of equal consideration for similar interests, about animals who cause harm to human beings?
- Should we try to stop a cat from killing a mouse, or a lion from killing a zebra?
- How do we know that plants cannot feel pain and, if they can, must we starve?

These are all good questions, but to avoid diverting the main argument of this book I am postponing my answers to chapter 6. If you are impatient to have your objections answered, I can't stop you looking there now. If you can wait, you will find that the next two chapters explore two examples of speciesism in practice.

I do not want to write a compendium of all the nasty things humans do to animals, so this book contains no discussion of practices like hunting for sport, the fur industry, keeping exotic pets, animals in circuses, or rodeos; and it says little about the impact of humans on wild animals. Instead I have gone more deeply into two central illustrations of speciesism in practice. These are not isolated examples of sadism but practices that involve in one case more than 100 million animals, and in the other well over

100 *billion* vertebrate animals every year. Nor can we pretend that we have nothing to do with these practices. One of them—experimentation on animals—is promoted by our governments and often paid for by our taxes. The other—raising animals for food—is possible only because most people buy and eat its products. These practices are the heart of speciesism. They cause more suffering to more animals than anything else that human beings do. To stop them we must change what we eat, and change the policies of our governments as well. If these officially promoted forms of speciesism can be stopped, abolition of the other speciesist practices will not be far behind.

CHAPTER 2

Tools for Research . . .

no, it's not all about saving human lives

ABOUT THIS CHAPTER

EACH YEAR, RESEARCHERS CONDUCT EXPERIMENTS THAT INFLICT pain and death on tens of millions of animals. Although this practice is usually defended on the grounds that the experiments contribute to progress in treating major diseases, this chapter shows that many of these experiments have no such aim, and no prospect of leading to important benefits for anyone other than the individuals or corporations conducting the experiments.

No one knows—not even approximately—how many animals are used, worldwide, in experiments. That's partly because, in the United States, the official regulatory body for the use of animals in experiments, the Department of Agriculture (USDA), does not collect data on how many animals are used. The reason for this surprising omission is that the U.S. Animal Welfare Act, which gives USDA authority to regulate research on animals, specifically excludes rats, mice, and birds—which make up the vast majority of animals used in experiments. Hence USDA has the authority to regulate experiments on, and to collect statistics about, only a small fraction of the animals actually experimented upon. It does collect figures on other mammals, including monkeys, dogs, and rabbits. So we know that in 2019, 797,546 animals other than rats, mice, and birds were used. That figure includes 18,270 cats, 58,511 dogs, and 68,257 nonhuman primates. But what proportion of all animals used in experiments are included in the USDA statistics? To find out, Larry Carbone, a veterinarian with four decades of experience working with laboratory animals at major universities, obtained data from eleven public and five private research facilities that were all among the top thirty facilities funded by the U.S. National Institutes of Health. The data showed that the facilities used about 39,000 animals of the species that they were required to report to the USDA, and more than 5.5 million mice and rats. Mice and rats therefore made up 99.3 percent of the animals these facilities used, which meant that the USDA figures counted only 0.7 percent of warm-blooded mammals experimented upon in these sixteen facilities. If a similar ratio applied to all U.S. research facilities reporting animal use to USDA, Carbone calculated, more than 112 million animals were used in the U.S. in 2018.[1] In contrast, the National Association for Biomedical Research claims—though without providing specific data—that rats and mice make up 95 percent of the animals used. If that is correct, the total number of animals used would be approximately 15.6 million.[2]

If we accept the figure at the lower end of the range, then U.S. usage of animals in experiments is far below that of China. Zhiyan Consulting, a Chinese business consulting firm, states that in 2021 the demand for rats and mice for scientific purposes was 49.8 million, for rabbits 2.2 million, for nonhuman primates 129,000 and for dogs, 64,000, making a total of more than 52 million animals. Although it is not clear that all of the animals were actually used in experiments, it seems likely that the great majority of them were.[3] Japan, another large user, has no official statistics. According to a survey conducted by the Japanese Association for Laboratory Animal Science, in the year 2008–2009, laboratories reared or maintained just over 15 million animals, but the accuracy of the survey is uncertain, and the number might well have changed since then.[4] European Union statistics, which cover all vertebrate animals, show about 10.4 million used in laboratories in 2019, a figure that includes the United Kingdom, which was then still part of the Union, and also Norway, which reports data to the European Union although it is not a member.[5] Adding in estimates for other countries that use fewer animals, we reach a range of 100 million to 200 million used annually worldwide, with the higher figure based on Carbone's calculation for the U.S. and the lower figure based on the estimate of the National Association for Biomedical Research. Carbone's figure is an extrapolation from limited data, and so could be mistaken, but in the absence of more reliable data, I will conclude that approximately 200 million animals are used each year in experiments.

The following pages describe a few of these experiments. Doing the research for this chapter, both in the present edition of this book and for the earlier versions, has always been a deeply disturbing experience. Reading this chapter is unlikely to be a pleasant experience either. But if animals have to undergo these experiments, the least we can do is inform ourselves about them, especially as, in many cases, our taxes are paying for them. I have avoided lurid language, but I have not attempted to tone down or gloss over experiments that are carried out on animals. Many of the accounts that follow are drawn from publications written by the experimenters themselves and published by peer-reviewed scientific journals. Such accounts are generally more favorable to the experimenters than reports by an outside observer would be. The experimenters will not emphasize the suffering they have inflicted unless it is necessary to do so in order to communicate the nature and results of the experiment. Most suffering therefore goes unreported. In addition, experimenters are unlikely to include in their publications any mention of something that goes wrong and harms an animal, especially if it is outside the context of the experiment itself. These things are usually discovered only when animal advocates go undercover into labs and record what is happening.

A second reason why scientific journals are a source favorable to experimenters is that they include only those experiments that the experimenters and editors of the journals judge significant. Researchers in many fields have become increasingly concerned about "publication bias"—the fact that journals tend to publish reports of experiments that achieve positive results. This bias makes published research findings unreliable because if five research groups independently test similar treatments and only one gets a positive result, that is likely to be the only one that is published, and the result will gain more credibility than it deserves. To ascertain the extent of this publication bias in studies using animals, a team led by Mira van der Naald, a researcher in cardiology at Utrecht University in the Netherlands, checked research applications involving animals at the University Medical Centre in Utrecht and linked these applications to publications during the seven years following the application. The team found that the publications described work involving only 26 percent of the animals, or 1,471 out of 5,590, which means that the research using almost three-quarters of the animals experimented upon was not published.[6] Utrecht University is the Netherlands' top-ranking university and ranks fifty-second in the world. Researchers at high-ranking universities are likely to be more successful in publishing their research than researchers at universities that rank lower, so it is likely that at most other institutions projects involving an even higher proportion of animals go unpublished.[7] In reading the following pages, therefore, if the results of the experiments do not appear to be of sufficient importance to justify the suffering they inflicted on animals, remember that these examples are all taken from the tip of the iceberg—the experimental results that journal editors considered significant enough to publish.

For readers who want to check my claims, I provide references to the original place of publication, including the name of the lead author. It should not be assumed that the authors are especially evil or cruel people. They are doing what they were trained to do and what thousands of their colleagues do. I describe the experiments to demonstrate the institutionalized mentality of speciesism that enables researchers to design and carry out experiments that are possible only if we ignore or discount the suffering of the animal subjects.

MAKING MONKEY PSYCHOPATHS

IN 2015 FOUR MEMBERS OF THE U.S. CONGRESS WROTE TO DR. Francis Collins, director of the National Institutes of Health (NIH), expressing concerns about a chain of experiments conducted by Dr. Stephen Suomi with the stated aim of producing monkey models of mental illness. According to a report published by People for the Ethical Treatment of Animals (PETA) Suomi had, over three decades and with funding from the NIH, deprived hundreds of infant monkeys of contact with their mothers, isolated them in small metal cages, and deliberately caused them to suffer anxiety, depression, diarrhea, hair loss, and to engage in forms of self-mutilation such as biting themselves and pulling out their own hair—social, emotional, and physical harm that lasted through their lives. Videos obtained by PETA show infant monkeys placed in tiny "startle chambers" in which they cannot even stand or crouch. Experimenters then frighten them with loud noises. The terrified monkeys cry out and try to escape, but there is no possibility of escape. Other videos show monkeys in small cages being repeatedly frightened by the presence of a human. They cry out and huddle at the back of the cage, trying to get as far as possible from the human. In another video, the infants are allowed to be with their mothers, but the mothers are given a drug that puts them to sleep and their nipples are taped over, so that the monkeys cannot suckle. A frightened infant is shown trying desperately to rouse their mother, as a young child might if they found their mother unresponsive.

Jane Goodall, who pioneered the study of free-living chimpanzees, said she was "shocked and saddened" by the videos. John Gluck, a researcher who had previously conducted experiments on the effect of depriving monkeys of social contact, later came to see that deprived monkeys are an inadequate model for human mental illness. He regarded Suomi's experiments, along with his own earlier work, as unjustified because of "the costs in suffering and pain." Barbara King, an expert in monkey and ape behavior, wrote in *Scientific American* about the importance of the warmth and protection of a mother for free-living infant monkeys. Then she added, in reference to Suomi's experiments:

> As a person who watches two beloved family members struggle with mental illness, I know the importance of research in this area. Yet systematic reviews tell us conclusively that animal models do not translate well to human mental health. To treat mental illness in humans requires

direct attention to the real stressors we experience in our own lives—not artificial ones that we make rhesus infants endure.[8]

After a year of PETA campaigning, the NIH announced that Suomi's laboratory would be closed and he would not be involved in further experiments on animals.[9] This was a rare instance in which public opinion halted the infliction of severe suffering on animals that served no important human purpose—but only after experiments of this kind had gone on for fifty years. And as this chapter will show, other animal experiments as bad as Suomi's are continuing in the United States and other countries.

Suomi was not an isolated "rogue experimenter," and he did not invent the idea of depriving infant monkeys of contact with their mothers. He studied under Harry Harlow, who had established a laboratory for behavioral research using primates. Harlow, who worked at the National Primate Research Center in Madison, Wisconsin, was for many years the editor of a leading psychology journal and, until his death in 1981, was held in high esteem by his colleagues in psychological research. His research on the effects of maternal deprivation in monkeys has been cited approvingly in many basic textbooks of psychology and read by millions of students taking introductory psychology courses.

In a 1965 paper, Harlow describes his work as follows:

> For the past ten years we have studied the effects of partial social isolation by raising monkeys from birth onwards in bare wire cages. These monkeys suffer total maternal deprivation. More recently we have initiated a series of studies on the effects of total social isolation by rearing monkeys from a few hours after birth until 3, 6, or 12

> months of age in [a] stainless steel chamber. During the prescribed sentence in this apparatus the monkey has no contact with any animal, human or sub-human.

These studies, Harlow continues, found that "sufficiently severe and enduring early isolation reduces these animals to a social-emotional level in which the primary social responsiveness is fear."[10]

Suomi continued to work with Harlow after graduating in 1971. In one study, he and Harlow describe how they were trying to induce psychopathology in infant monkeys by a technique that appeared not to be working. They were then visited by John Bowlby, a British psychiatrist, who listened to the story of their troubles and toured the Wisconsin laboratory. After he had seen the monkeys individually housed in bare wire cages he asked, "Why are you trying to produce psychopathology in monkeys? You already have more psychopathological monkeys in the laboratory than have ever been seen on the face of the earth."[11]

Harlow is sometimes credited with discovering the serious psychological consequences of maternal deprivation, but Bowlby had, several years earlier, already demonstrated these consequences through research conducted with children who had been deprived of their mothers, primarily war orphans, refugees, and institutionalized children. As far back as 1951, Bowlby concluded:

> The evidence has been reviewed. . . . It is submitted that evidence is now such that it leaves no room for doubt regarding the general proposition—that the prolonged deprivation of the young child of maternal care may have grave and far-reaching effects on his character and so on the whole of his future life.[12]

This did not deter Harlow and his colleagues from devising and carrying out experiments from the late 1950s onward, that deprived infant monkeys of their mothers and then, in the following decades, to use even more extreme forms of deprivation in an attempt to create monkey models of depression and psychopathology. In fact, in the same article in which they tell of Bowlby's visit, Harlow and Suomi describe how they had the "fascinating idea" of inducing depression by "allowing baby monkeys to attach to cloth surrogate mothers who could become monsters":

> The first of these monsters was a cloth monkey mother who, upon schedule or demand, would eject high-pressure compressed air. It would blow the animal's skin practically off its body. What did the baby monkey do? It simply clung tighter and tighter to the mother, because a frightened infant clings to its mother at all costs. We did not achieve any psychopathology.
>
> However, we did not give up. We built another surrogate monster mother that would rock so violently that the baby's head and teeth would rattle. All the baby did was cling tighter and tighter to the surrogate. The third monster we built had an embedded wire frame within its body which would spring forward and eject the infant from its ventral surface. The infant would subsequently pick itself off the floor, wait for the frame to return into the cloth body, and then cling again to the surrogate. Finally, we built our porcupine mother. On command, this mother would eject sharp brass spikes over all of the ventral surface of its body. Although the infants were distressed by these pointed rebuffs, they simply waited until the spikes receded and then returned and clung to the mother.

These results, the experimenters remark, were not so surprising, since the only recourse of an injured child is to cling to its mother.

Eventually, Harlow and Suomi gave up on the artificial monster mothers because they found something they liked better: a real monkey mother who was a monster. To produce such mothers, they reared female monkeys in isolation, then tried to make them pregnant. These females would not have normal sexual relations with male monkeys, so they had to be made pregnant by a technique that Harlow and Suomi referred to as a "rape rack." When the babies were born the experimenters observed the monkeys. They found that some simply ignored the infants, failing to cuddle the crying baby to the breast as normal monkeys do when they hear their baby cry. The other pattern of behavior observed was different:

> The other monkeys were brutal or lethal. One of their favorite tricks was to crush the infant's skull with their teeth. But the really sickening behavior pattern was that of smashing the infant's face to the floor, and then rubbing it back and forth.[13]

In a 1972 paper, Harlow and Suomi say that because depression in humans has been characterized as embodying a state of "helplessness and hopelessness, sunken in a well of despair," they designed a device "on an intuitive basis" to reproduce such a "well of despair" both physically and psychologically. They built a vertical chamber with stainless steel sides sloping inward to form a rounded bottom and placed a young monkey in it for periods of up to forty-five days. They found that after a few days of this confinement the monkeys "spend most of their time huddled in a corner of the chamber." The confinement produced "severe and persistent psychopathological behavior of a depressive nature." Even nine months after release the monkeys would sit clasping their arms around their bodies instead of moving around and exploring their surroundings as normal monkeys do. But the report ends inconclusively and ominously:

> Whether [the results] can be traced specifically to variables such as chamber shape, chamber size, duration of confinement, age at time of confinement or, more likely, to a combination of these and other variables remains the subject of further research.[14]

Another paper explains how, in addition to the "well of despair," Harlow and his colleagues created a "tunnel of terror" to produce terrified monkeys,[15] and in yet another report Harlow describes how he was able "to induce psychological death in rhesus monkeys" by providing them with terry cloth–covered "mother surrogates" that were normally kept at a temperature of 99° Fahrenheit (37°C) but could be rapidly chilled to 35° Fahrenheit (1.5°C) to simulate a kind of maternal rejection.[16]

After Harlow's death, his students continued to perform experiments in a similar vein. One of them, Gene Sackett, continued deprivation studies at the University of Washington Primate Center, only to find that his research raised doubts about the purported justification of all this work, namely its value for treating psychopathology in humans. Sackett reared infants of three different species of monkey—rhesus monkeys, pigtail macaques, and crab-eating macaques—in total isolation to study the differences in personal behavior, social behavior, and exploration behavior. He found that the species differ in ways that "question the generality of the 'isolation syndrome' across primate species."[17] If there are differences among species of monkeys, generalization from monkeys to humans must be far more questionable.

Deborah Snyder of the University of Colorado conducted deprivation experiments on bonnet monkeys and pigtailed macaques. She was aware that Jane Goodall's observations of orphaned wild chimpanzees described "profound behavioral disturbances, with sadness or depressive affective changes as major components." But because "in comparison with monkey studies, relatively little has been published on experimental separations in great apes," she and other experimenters decided to study seven infant chimpanzees who had been separated from their mothers at birth and reared in a nursery environment. After periods ranging between seven and ten months, some of the infants were placed in isolation chambers for five days. The isolated infants screamed, rocked, and threw themselves at the walls of the chamber. The researchers concluded that "isolation in infant chimpanzees may be accompanied by marked behavioral changes" but predictably noted that more research was needed.[18]

Since Harlow began his maternal deprivation experiments more than sixty years ago, hundreds of similar experiments have been conducted in the United States and elsewhere, subjecting thousands of animals to procedures that induce distress, despair, anxiety, general psychological devastation, and death. As some of the preceding quotations show, this research feeds off itself. Snyder and her colleagues experimented on chimpanzees just because relatively little experimental work had been done on the great apes, as compared with monkeys. They apparently felt no need to address the basic question of why we should be doing any experiments on maternal deprivation in animals at all. They did not even try to justify their experiments by claiming they were beneficial to human beings. Their attitude was: this has been done with animals of various species, but not with animals of this particular species, so let's do it to them. They took this stance even though it must have been obvious to them that if species differ in their response to maternal deprivation, *Homo sapiens* is likely to differ as well. Similar rationales for harming yet more animals occur frequently throughout the psychological and behavioral sciences.

PSYCHOLOGY'S ETHICAL DILEMMA

RESEARCHERS IN PSYCHOLOGY WHO USE ANIMALS FACE AN ETHICAL dilemma: Either the minds of the animals are not like ours, in which case the experiments are unlikely to benefit us and there is less justification for funding and carrying them out; or else the animals do have minds like ours, in which case we ought not to perform on the animal an experiment that would be considered outrageous if performed on one of us. That so many researchers in psychology who perform extremely painful experiments on animals manage not to see this obvious problem is further confirmation of Upton Sinclair's remark that "It is difficult to get a man to understand something when his salary depends on his not understanding it."[19]

It isn't easy to estimate how many animals are experimented on in psychology research, or in other research that is supposedly going to help prevent or treat mental illness. A search of the NIH *RePORTER* found that in 2020 the U.S. National Institute of Mental Health provided $185 million for 304 research projects involving experimentation on animals, but this does not tell us how many animals were used in each research project, nor does it include research funded from other sources.[20]

LEARNED HELPLESSNESS

ANOTHER LONG AND DISTURBING LINE OF EXPERIMENTS IN PSYchology goes under the heading "learned helplessness," supposedly to attempt to model depression in human beings. In 1953, R. Solomon, L. Kamin, and L. Wynne, experimenters at Harvard University, placed forty dogs in a device called a shuttlebox, which consists of a box divided into two compartments separated by a barrier. Initially the barrier was set at the height of the dog's back. Hundreds of intense electric shocks were delivered to the dog's feet through a grid floor. At first the dog could escape the shock if he learned to jump the barrier into the other compartment. In an attempt to "discourage" one dog from jumping, the experimenters forced that dog to jump one hundred times onto a grid floor in the other compartment that also delivered a shock to the dog's feet. They said that as the dog jumped he gave a "sharp anticipatory yip which turned into a yelp when he landed on the electrified grid." They then blocked the passage between the compartments with a piece of plate glass and tested the dog again. The dog "jumped forward and smashed his head against the glass." The dogs began by showing symptoms such as "defecation, urination, yelping and shrieking, trembling, attacking the apparatus," and so on; but after ten or twelve days of trials, dogs who were prevented from escaping shock ceased to resist. The experimenters reported themselves "impressed" by this and concluded that a combination of the plate glass barrier and foot shock was "very effective" in eliminating jumping by dogs.[21]

This study showed that it was possible to induce a state of hopelessness and despair by repeated administration of severe inescapable shock. In the 1960s other experimenters took up the idea and further refined the study of "learned helplessness." Prominent among them was Martin Seligman of the University of Pennsylvania. In one study, written with colleagues Steven Maier and James Geer, Seligman describes his work as follows:

> When a normal, naive dog receives escape/avoidance training in a shuttlebox, the following behavior typically occurs: at the onset of electric shock the dog runs frantically about, defecating, urinating, and howling until it scrambles over the barrier and so escapes from shock. On the next trial the dog, running and howling, crosses the barrier more quickly, and so on, until efficient avoidance emerges.[22]

Seligman and his colleagues altered this pattern by strapping dogs into harnesses and giving them shocks from which they had no means of escape. When the dogs were then placed in the original shuttlebox situation from which escape was possible, they found that:

> such a dog reacts initially to shock in the shuttlebox in the same manner as the naive dog. However, in dramatic contrast to the naive dog, it soon stops running and remains silent until shock terminates. The dog does not cross the barrier and escape from shock. Rather it seems to "give up" and passively "accept" the shock. On succeeding trials the dog continues to fail to make escape movements and thus takes 50 seconds of severe, pulsating shock on each trial. A dog previously exposed to inescapable shock may take unlimited shock without escaping or avoiding at all.[23]

Subsequently Seligman completely changed the focus of his research and became well known for his work in "positive psychology"—the study of what makes life most worth living. For many years, I wondered if he had, belatedly, seen the ethical dilemma inherent in causing so much distress to animals and so had changed course professionally. I even hoped that he would express some remorse for what he had done and discourage others from doing it too. But then I came across "Learned Helplessness at Fifty: Insights from Neuroscience," an article he co-authored in 2016 with Steven Maier. There I found this paragraph:

> We must mention that running dog experiments was a harrowing experience for both of us. We are both dog lovers and as soon as we could we stopped experimenting with dogs and used rats, mice, and people in helplessness experiments, with exactly the same pattern of results.[24]

Rats can feel pain too—if they didn't, how could you shock them into a state of learned helplessness?—and if you have read the first chapter of this book, you will know that I don't regard the fact that we have less love for rats and mice than we have for dogs as a good reason for not giving equal consideration to their suffering.

Perhaps because of increased public sensitivity to the use of dogs, other researchers on learned helplessness also switched to using other animals. Dogs, rats, mice, and even goldfish were used in learned helplessness experiments, but with mixed results. At the University of Tennessee at Martin, G. Brown, P. Smith, and R. Peters created a specially designed shuttlebox in which they subjected forty-five fish to sixty-five shock sessions each, only to conclude that "the data in the present study do not provide much support for Seligman's hypothesis that helplessness is learned."[25]

Steven Maier made a career out of inflicting learned helplessness in animals. Yet in 1984 he had this to say about comparing such animals with depressed humans:

> It can be argued that there is not enough agreement about the characteristics, neurobiology, induction, and prevention/cure of depression to make such comparison meaningful. . . . It would thus appear unlikely that learned helplessness is a model of depression in any general sense.[26]

Although Maier tried to salvage something from this dismaying conclusion by saying that learned helplessness may constitute a model not of depression but of "stress and coping," he had effectively admitted that more than thirty years of animal experimentation had been a waste of time and of substantial amounts of taxpayers' money, quite apart from the immense amount of acute physical pain caused by such experiments.

When Maier acknowledged that learned helplessness was not a suitable animal model for depression, you might imagine that funding for research that involves shocking animals into a state of learned helplessness would dry up. You would be wrong. In 1992 Hielke van Dijken and other Dutch researchers described how giving inescapable shocks to rats' feet had a long-lasting impact on their subsequent behavior when exposed to a change in their environment. They suggested that this kind of prior stress might serve as an animal model for investigating post-traumatic stress disorder (PTSD) and other conditions.[27] Now researchers had a new reason for subjecting animals to repeated inescapable electric shock: to provide a model of PTSD. It is a reason they are still using today.

In a 2019 paper titled "Modelling post-traumatic stress disorders in animals," Bibiana Török and colleagues from Hungary and Croatia describe the wide variety of methods used by researchers to create an animal model for PTSD.[28] The most frequently used method, they report, is the "single prolonged stress paradigm," which involves subjecting the animals to a sequence of four different stresses. A typical example might go like this: First is "restraint"—two hours in a tube that prevents all movement. Next is the "forced swim test"—the animals are put in deep water from which they cannot escape and can only avoid drowning by swimming for twenty minutes. Third, they are put into a chamber and exposed to ether, which causes them to lose consciousness. Finally, inescapable electric shock is delivered to their feet through an electrified grid floor for up to thirty seconds, or in some cases, shorter shocks are repeated at intervals for ninety minutes. Other stress models include:

- footshock alone;
- exposing an animal to a predator, or to the odor of a predator;
- prolonged and severe immobilization, sometimes combined with exposure to predator odor;
- "social defeat," which involves forced contact with a trained, aggressive animal of the same species, followed by being put in close proximity to the aggressive animal but not in physical contact with it;
- "underwater trauma"—a forced swim followed by being held underwater for twenty seconds.

All of these techniques have been used by the researchers cited by Török and colleagues. Yet Török and others question whether such techniques can produce adequate models of PTSD. Nonhuman animals cannot tell us what they are feeling after being subjected to these painful and stressful experiences, so researchers try in various ways to assess whether the animals are experiencing something like PTSD. Researchers use a variety of tests, such as the levels of specific biomarkers in blood samples, behavior that may indicate a negative mood, disrupted sleep, reduced social interaction, and an increased "startle response." The percentage of animals in which these different forms of trauma leads to such indications of something like PTSD varies with the species of animal used and even with particular breeds of animal within a species. In one strain of rats, known as Sprague Dawley rats, Török's team reports that 30 to 50 percent of the animals could be considered as vulnerable, a figure significantly higher than in humans subjected to trauma, only 10 to 20 percent of whom subsequently develop PTSD.

To explain this low percentage of PTSD in humans who have experienced trauma, some researchers have developed what is known as the "three hit theory": that those who suffer from PTSD have genes that predispose them to it, experience adversity early in life, and then face a particular challenge. The various stresses just described are attempts to model only the third of these "hits." Török's team notes this problem with past attempts to produce PTSD in animals, but instead of abandoning the attempt to produce an animal model, they advocate seeking to reproduce all three of these factors. To ensure that genetic variants in the population can be studied, they recommend that "researchers should focus on coping strategies of large populations"—which implies that every individual animal in a large population should be stressed. Because stresses in childhood may be implicated in susceptibility to PTSD, they also recommend mimicking such early-life stressors by separating the animals from their mothers for three hours a day on each of the first ten days after birth, or forcing young animals to keep swimming for prolonged periods. This should then by followed by the infliction of a traumatic event in adulthood.

In other words, although these researchers acknowledge that their predecessors have failed to produce a useful animal model, they want to try again, with more stress for more animals and over a longer period. They suggest this despite having themselves given an account of the likely causes of PTSD in humans that should have led them to question the very possibility of achieving a useful animal model. They write that in humans, in addition to genetic risk factors, it is also necessary to take into account, among other factors, "the childhood home life and years of education . . . as pre-stress variables and post-trauma life stress and post-trauma social support. . . ." One can only wonder how researchers will incorporate childhood home life and a varying number of years of education into their animal models.

Sadly, despite all the suffering and the doubts about the applicability of creating animal models of PTSD, this seems to be a booming field. Lei Zhang and a team of colleagues searched PubMed, a major index of medical research, and found 792 publications on animal models of PTSD prior to August 2018, most of them from the period 2007 to 2018. Like Török, they found models created by stresses that included "learned helplessness, footshock, restraint stress, inescapable tail shock, single-prolonged stress, underwater trauma, social isolation, social defeat, early life stress, and predator-based stress." The animals used included mice, rats, pigs, birds, monkeys, dogs, and cats.[29]

Meanwhile, other researchers are still seeking to create animal models of depression, varying the species of animal used and the forms of stress. Meghan Donovan, Yan Liu, and Zuoxin Wang, all of Florida State University, put prairie voles (small rodents native to American grasslands) in plastic tubes and used plastic mesh and Velcro straps to, in their own words, "completely immobilize the subject" and kept them there for a full hour. They did this because this kind of immobilization had been found, in previous research, to cause stress to the voles. (No surprise there.) But why use prairie voles? Because, the researchers say, "Humans, as social creatures, have consistently relied on social bonds to ensure survival and success," and the presence of these bonds can protect against anxiety, depression, and other negative states. Previous attempts to model these protective effects in animals have failed because most animal models lack "the complex social structures seen in humans." The researchers then point out that prairie voles are predominantly monogamous and form pair-bonds. Their study showed that the presence of a partner reduced the signs of stress in the immobilized vole. Research using consenting human subjects might have shown something similar, but as the voles were decapitated immediately after the testing was completed and their brains sectioned, that aspect of the study would not have been possible. The researchers conclude that "As social environments are a critical part of our lives, we must continue to explore this area of research to understand how social bonds may ultimately shape our health outcomes and well-being."

Voles may resemble humans in being predominantly monogamous, but their monogamy is not an adequate reason to subject them to an hour of severe stress—and if vole pair-bonds really are anything like human relationships, the partners observing the immobilized voles must also be undergoing an extremely stressful experience. This research was funded by the National Institutes of Health and is only one of several experiments involving stressed voles by various authors. The social environments of voles are very different from ours, and this seems likely to be yet another animal model that will eventually, after the infliction of great suffering and the expenditure of millions of dollars of public funds, be rejected and abandoned.[30]

In an example of research that took place in China but with the involvement of U.S. researchers, a team led by T. Teng included Carol Shively of Wake Forest University in North Carolina and Gretchen Neigh of Virginia Commonwealth University, as well as other researchers from Chongqing Medical University. The researchers reared monkeys in large social groups for the first year of their lives, then put them into cages, alone, for seventy days. They were then subjected to two different forms of stress per day, in five cycles of seven days each. After each cycle they then had a four-day interval in which measurements were taken before the next stress cycle began. On each of the thirty-five days on which they were stressed, these isolated adolescent monkeys were subjected to two of the following: twelve hours of loud noise; no water for twelve hours; no food for twenty-four hours; a restricted space for four hours; twelve hours of a stroboscope; cold temperatures of 10°C (50°F) for ten minutes; more restricted space for four hours; and inescapable electric shocks delivered to their feet for ten to fifteen seconds, three or four times. Despite the title of the paper suggesting that this was only "mild" stress, it was severe enough to change the monkeys' behavior. Compared to the control group, they spent significantly more time in what the researchers called "huddle posture"—monkeys hanging their heads down to shoulder level or lower and clasping themselves. The researchers conclude that the stresses they inflicted "can induce depressive-like and anxiety-like behaviors."[31]

Elsewhere in China, researchers are using similar methods to make monkeys depressed. I will give two more examples.

Weixin Yan and a large team of researchers, mostly from Southern Medical University in Guangzhou, housed ten monkeys in individual cages with a floor area just 42 cm by 30 cm and a height of 50 cm. In a preprint—a version of the article posted online when it has not yet been accepted by a journal—the researchers described this degree of space restriction thus: "Their activities are limited, barely able to turn around. . . . They cannot see and touch each other."

While in this extreme form of confinement, the monkeys were subjected to ten different stressors. The researchers wrote that the stressors were "randomly arranged throughout the day and night over 90 consecutive days. The stressors were (1) food deprivation, (2) water deprivation, (3) overnight illumination, (4) all-day light deprivation, (5) ice-water, (6) social isolation, (7) restraint, (8) stick stimulation, (9) stick cage and (10) cage dystopia." The meaning of some of these stressors is unclear, and the preprint does not provide further details. The experiment was subsequently published in a leading medical journal and then retracted. In retracting the article, the journal referred to "the shortcomings of the cynomolgus monkey depression model," and added that "The isolated conditions applied, and the flawed behavioural classification method and the insufficient experimental duration of the animal model render the conclusions of the article inaccurate." In other words, ten monkeys endured 90 days of extreme suffering for nothing.[32]

For eight weeks, Yong-Yu Yin of the Beijing Institute of Pharmacology and Toxicology and other researchers in China subjected ten monkeys to two forms of stress every day, carried out according to the following plan:

THE STRESSORS AND ROUTINES OF CHRONIC UNPREDICTABLE STRESS IN THE FIRST WEEK

Day	Stressor 1 (Day 7:30–19:30)	Stressor 2 (Night 19:30–7:30)
Monday	Food deprivation	Strobe light
Tuesday	Water deprivation	Loud noise
Wednesday	Intimidation	Space restriction
Thursday	Loud noise	Food deprivation
Friday	Space restriction	Strobe light
Saturday	Intimidation	Loud noise
Sunday	Water deprivation	Space restriction

According to the research paper, "Two stressors were applied each day, and each stressor generally lasted 12 hours." Intimidation was done using fake snakes, which monkeys fear.[33]

People for the Ethical Treatment of Animals asked Emily Trunnell, who has a doctorate in neuroscience from the University of Georgia, to review the three experiments I have just described. "Essentially," she wrote, "monkeys were tortured for these experiments." The stresses inflicted on the monkeys, however, were quite different from those that commonly cause depression or other forms of mental illness in humans, such as sexual and physical abuse, relationship problems, financial difficulties, addiction, illness or injury, and various combinations of these problems. Hence Trunnell concluded that a nonhuman primate animal model is unlikely to be applicable to human depression.[34]

We have now looked at psychology experiments attempting by various methods to produce animal models of depression and post-traumatic stress disorder. All of them have inflicted prolonged and often severe suffering on animals. The animals who suffered in these types of research are just a few of the vastly larger number of animal subjects who suffer each year in psychological experiments. The data suffices, though, to show that the misery and pain inflicted on animals that I described in the original 1975 edition of this book has not ceased. Many experiments are still being conducted that cause great pain to animals yet offer little prospect of yielding vital new knowledge. It is still common for experimenters to treat animals as mere tools for research. Researchers today may be more careful about their public image than Harlow and Suomi were when they described their method of making an unwilling female monkey pregnant by means of a "rape rack." But behind the better public face, many researchers working in psychology are still treating animals as if their suffering doesn't really count.

POISONING, BLINDING, AND OTHER WAYS OF TESTING ON ANIMALS

ANOTHER MAJOR FIELD OF EXPERIMENTATION ON ANIMALS involves the poisoning of millions of animals annually in compliance with regulations for testing substances on animals before they are released for human use. In the European Union in 2019, 1,788,779 "uses of animals" were for testing substances in accordance with regulations.[35] The researchers classified 9 percent of these as involving "severe pain" and 33 percent as involving "moderate pain," meaning that in that year approximately 750,000 procedures were conducted that caused more than just "mild pain" and that, of these, about 160,000 involved severe pain—and that is assuming that the researchers accurately report how much pain their experiments caused.[36] In 2019 in Great Britain, 437,124 scientific procedures were performed on animals to test substances in accordance with regulations. Of these, researchers reported 80,371 as involving moderate pain, and 47,574 as involving severe pain.[37]

In the United States, as we have seen, only a tiny fraction of the animals used in research are subject to government regulation, and even for those animals, such as dogs, cats, and monkeys, researchers are not required to assess the severity of the pain they inflict on animals. Therefore we can only guess at the numbers of animals used for specific purposes, including regulatory testing, and the number involving severe pain. In the European Union, 17 percent of all uses of animals for scientific purposes in 2019 were for regulatory testing,[38] while in Great Britain in 2019 the corresponding figure was 25 percent. If the proportion in the United States is similar, let's say 21 percent, and if Carbone's estimate of a total of 112 million animals used annually in regulatory research is correct, the number of animals used in regulatory testing must be around 24 million. Of those 24 million animals used for regulatory testing, if we again take a midpoint between the figures for the European Union and for Great Britain, then 10 percent, or 2.4 million, will have undergone severe pain. The true figures could easily be double or triple those numbers, because there is so much more research and development in this field in the United States than in Europe or Britain, and the Food and Drug Administration requires extensive testing of new substances before they are released. It may be thought justifiable to require tests on animals of potentially life-saving drugs, but the same kinds of tests are used for products like food coloring and floor cleaners. For example, in the European Union, 61 percent of regulatory testing is for medical products and another 18 percent is for veterinary products, with the largest remaining category being "industrial chemicals," involving 154,397 uses of animals. Should thousands of animals suffer so that new kinds of food coloring or cleaning products can be put on the market?

Even when animals are used to test a medical product, the product may not improve our health. Pharmaceutical companies spend large sums on producing "me-too" drugs that they hope will gain a share of a lucrative market dominated by a drug patented by a rival company. The drugs they produce gain approval from regulatory authorities by being compared to a placebo, not to drugs already available, so they do not need to be an improvement on the rival drug—and they may even be worse. Jeffrey Aronson and Richard Green, of the Universities of Oxford and Nottingham, respectively, examined the World Health Organization's list of essential drugs and classified 60 percent of them as me-too drugs, including popular brands of statins, such as Lipitor. Although they acknowledge that some me-too drugs may have advantages over the original drugs with which they compete, they found that many of them offer "no significant advantages over their predecessors." As an example, they cite angiotensin-converting enzyme (ACE) inhibitors, drugs that lower blood pressure. They found fifteen different ACE inhibitors listed on national essential medicine lists. Only the first two of these introduced significant benefits, they report, with none of the remaining thirteen, including drugs such as lisinopril (sold as Prinivil and Zestril), having "any important advantages."[39]

In order to determine how poisonous a substance is, "acute oral toxicity tests" are performed. These tests were developed in the 1920s to provide a standard for comparing how poisonous various substances are. Up to a hundred animals would be used, divided into groups receiving different dosages. The standard chosen for comparing the toxicity of different substances was the quantity of the substance ingested (milligrams per kilogram body weight) at which half the animals died. This came to be known as the "lethal dose 50 percent," or LD50, test. Over the following decades, determining the LD50 of a substance became not just a method used by scientists seeking to compare the toxicity of different substances but a benchmark that corporations had to meet before any new product could be put on the market. Moreover, the tests typically had to be performed on animals of at least two different species. By the 1980s, in the United States alone, as a result of the use of this benchmark, the number of animals poisoned to death each year had risen into the millions.[40]

Many animals are good at detecting and avoiding a wide range of poisons in their food—they have to be, to survive in the wild. When animals will not eat the substance researchers wish to test, experimenters force-feed them by inserting a tube down their throats, a technique known as gavage and notorious for its use on ducks and geese to create foie gras. One study set out to investigate the health effects of repeated gavage on mice. The person administering the gavage was, the researchers reported, an experienced technician who used special handling techniques to reduce stress and anxiety, and the gavage was carried out only when the animals were calm. It is reasonable to suppose that not all technicians would be equally experienced or would go to such trouble to avoid stressing the animals. Yet in the group receiving gavage, the mortality rate was 15 percent. The dead mice had perforations in their esophagus, indicating that even this experienced technician, taking care to calm the mice, could not reliably avoid causing fatal, and presumably very painful, injuries. In the control group, kept in identical conditions but without gavage, no mice died.[41]

Standard tests involve a daily dose of the substance being tested for fourteen days but some may last for up to one year—for those animals who survive that long. Normally, before reaching the point at which half of them die, they will all or nearly all become very ill and will be in obvious distress. In the case of fairly harmless substances researchers may still seek to ascertain the concentration that will make half the animals die; consequently enormous quantities have to be force-fed to the animals, and death may be caused merely by the large volume or high concentration given to the animals, even though these concentrations have no relevance to the circumstances in which humans will use the product. During this time, the animals often display classic symptoms of poisoning, including vomiting, diarrhea, paralysis, convulsions, and internal bleeding. Since the very point of these experiments is to measure how much of the substance will poison half the animals to death, dying animals are not put out of their misery for fear of producing inaccurate results.

In addition to acute toxicity testing, many substances are

tested for potential eye irritancy by being placed in the eyes of rabbits, used because of their large eyes, because they are easy to handle, and because they are cheap. Here too, for many decades, there was a standard test: the Draize eye-irritancy test, developed in the 1940s, when J. H. Draize, working for the U.S. Food and Drug Administration, developed a scale for assessing how irritating a substance is when placed in rabbits' eyes. To prevent them from scratching or rubbing their eyes, the animals were usually placed in holding devices from which only their heads protruded. A test substance was then placed in one eye of each rabbit. The method used was to pull out the lower eyelid and place the substance into the small "cup" thus formed. The eye was then held closed. Sometimes the application was repeated. The rabbits were observed daily for eye swelling, ulceration, infection, and bleeding. The studies lasted up to three weeks. The standard scale for measuring the response included a column headed "Pain Evaluation (Reaction by Animal)," in which the highest level of reaction was described as follows:

> Total loss of vision due to serious internal injury to cornea or internal structure. Animal holds eye shut urgently. May squeal, claw at eye, jump, and try to escape.[42]

Some substances cause such serious damage that the rabbits' eyes lose all distinguishing characteristics—the iris, pupil, and cornea begin to resemble one massive infection. In the form in which the Draize test was performed for many decades, experimenters were not obliged to use anesthetics. Though occasionally used when introducing the substance, the use of a topical anesthetic would have done nothing to alleviate the pain that can result over the three weeks following the placing of the substance in the eye.

Animals were, and in some circumstances still are, also subjected to other tests to determine the toxicity of many substances. Just as the LD50 seeks to find the lethal dose for 50 percent of the animals forced to ingest a substance, so the LC50 aims to discover the lethal concentration of sprays, gases, and vapors that animals in sealed chambers are forced to inhale. In dermal toxicity studies, rabbits have their fur removed so that a test substance can be placed on their skin. The animals are restrained so that they do not scratch at their irritated bodies. The skin may bleed, blister, and peel. Immersion studies, in which animals are placed in vats of diluted substances, sometimes cause the animals to drown before any test results can be obtained. In injection studies, the test substance is injected directly into the animal, either under the skin, into the muscles, or directly into an organ.

Toxicologists have known for a long time that extrapolation from one species to another is a highly risky venture. Oraflex, sold in the UK as Opren and touted by its manufacturer, the pharmaceutical giant Eli Lilly, as a new "wonder drug" for the treatment of arthritis, passed all the usual animal tests before it was released in 1980 in the UK. Over the next two years it caused 3,500 adverse reactions and 96 deaths. In the United States, where the drug was only released in May 1982, 43 deaths were reported before it was suspended from all markets in August 1982. Mice, it turned out, could tolerate far higher doses than humans with no ill effects.[43] Vioxx, another popular anti-arthritis drug that had been tested on animals, was first released in 1999 and taken off the market in 2004, but not before, according to U.S. Food and Drug Administration investigator David Graham, it had caused 60,000 deaths.[44] In a lawsuit against Merck, the manufacturer of Vioxx, the Physicians Committee for Responsible Medicine claimed that Merck relied on animal tests to justify continuing to sell Vioxx, even though studies with humans taking the drug had shown an increased risk of heart attacks.[45] In 2011 Merck pleaded guilty to a charge of illegal promotional activity in its marketing of Vioxx and agreed to pay $950 million to settle both civil and criminal actions against the company.[46]

As well as exposing people to harm, testing on animals may lead us to miss out on valuable products that are dangerous to animals but not to human beings. Insulin, for example, can produce deformities in infant rabbits and mice but not in humans.[47] Morphine, which is calming to human beings, causes mice to go into a frenzy. And as another toxicologist has said: "If penicillin had been judged by its toxicity on guinea pigs, it might never have been used on man."[48]

In the 1980s, after decades of mindless animal testing, some scientists began to have second thoughts. Dr. Elizabeth Whelan, a scientist and executive director of the American Council on Science and Health, pointed out: "It doesn't take a Ph.D. in the sciences to grasp the fact that rodent exposure to the saccharin equivalent of 1,800 bottles of soda pop a day doesn't relate well to our daily ingestion of a few glasses of the stuff." Whelan welcomed the fact that officials at the Environmental Protection Agency downgraded earlier estimates of risks of pesticides and other environmental chemicals, noting that the evaluation of cancer risks derived from animal extrapolation was based on "simplistic" assumptions that "strain credibility." This meant, she said, that "our regulators are beginning to take note of the scientific literature rejecting the infallibility of the laboratory animal test."[49] In a 2009 article in the *Journal of the Royal Society of Medicine*, Michael Bracken of Yale University Schools of Medicine and Public Health listed several examples of variations among species, including the fact that thalidomide does not cause birth defects in many species but does in humans, whereas corticosteroids cause birth defects in many nonhuman animals but not in humans.

He also notes that even testing on nonhuman primates may not predict a life-threatening reaction in humans.[50] Michael Schuler, a pioneer in biochemical engineering and professor emeritus at Cornell University, has quoted a major pharmaceutical company as telling him that "only 6 percent of animal trials are truly predictive of human response."[51]

By the 1980s, criticism of the use of animals in toxicity testing was also coming from activists in the growing animal rights movement. In the United States, Henry Spira, a civil rights activist who had marched alongside Blacks in the South to protest against segregation, used the Freedom of Information Act to obtain government documents filed by public institutions and major corporations describing experiments they were carrying out on animals. Then, armed with the information that Revlon, at that time the leader of the U.S. cosmetics industry, was testing its cosmetics by putting them into the eyes of rabbits, he went to that company and suggested that it put one hundredth of one percent of its gross revenues toward developing an alternative to the Draize test. When Revlon took no action, Spira gathered sufficient funds from his supporters to place a full-page advertisement in the *New York Times.* Under the headline "How Many Rabbits Does Revlon Blind for Beauty's Sake?"[52] readers saw a photo of a white rabbit with bandages over its eyes and glass laboratory flasks nearby. Beneath, the text read:

> Imagine someone placing your head in a stock. As you stare helplessly ahead, unable to defend yourself, your head is pulled back. Your lower eyelid is pulled away from your eyeball. Then chemicals are poured into the eye. There is pain. You scream and writhe hopelessly. There is no escape. This is the Draize Test. The test which measures the harmfulness of chemicals by the damage

> inflicted on the unprotected eyes of conscious rabbits. The test that Revlon and other cosmetic firms force on thousands of rabbits to test their products.

Roger Shelley, then Revlon's vice president for investor relations, said later: "I knew the stock was going down that day, but more importantly, I knew the company had a very significant problem that could affect not just one day's stock price, but could cut to the core of the company." Shelley was right. There were protests, covered on TV, outside Revlon's New York head office. People in rabbit costumes appeared at Revlon's annual meeting. Revlon got the message and allocated the requested funds to pay for research on alternatives to animal experiments. Other companies, such as Avon and Bristol-Myers, followed suit.[53] These funds helped to establish the Johns Hopkins Center for Alternatives to Animal Testing, in Baltimore.

Corporations such as Avon, Bristol-Myers, and Procter & Gamble began using alternatives in their own laboratories, thus reducing the number of animals used. In 1989 Avon announced that it had validated tests using a specially developed synthetic material called Eytex as a replacement for the Draize test, and subsequently both Avon and Revlon announced a permanent end to all animal testing.[54]

Although the most dramatic developments took place in the highly public and therefore relatively vulnerable cosmetics industry, the movement against animal testing was also having an impact on wider areas of industry. As a report in *Science* put it in 1987:

> Prodded by the animal welfare movement, major manufacturers of pharmaceuticals, pesticides and household products have made significant advances in recent years

> toward the goal of reducing the number of animals used in toxicity testing. Alternative methods, such as cell and tissue culture and computer modeling, are increasingly being seen not just as good public relations but as desirable both economically and scientifically.[55]

The report went on to quote Gary Flamm, then director of the Food and Drug Administration Office of Toxicology Sciences, as saying that the LD50 "should be replaceable in the vast majority of cases." A *New York Times* article quoted a senior toxicologist at G. D. Searle and Company as admitting that "an awful lot of the points made by the animal welfare movement are extreme but right."[56]

The toxicologist may have had in mind any number of callous, stupid things that were done just because regulations required them. It is hard to escape the conclusion that until there was an active animal rights movement, those in charge of animal testing regulations did not really think about animal suffering. How else could one explain the fact that, for example, it was not until 1983 that U.S. federal agencies were willing to state that substances known to be caustic irritants, such as lye, ammonia, and oven cleaners, did not need to be tested on the eyes of conscious rabbits?[57]

By the early 1990s it seemed that pressure from the animal movement, together with advances in the development of alternative methods of testing, was saving many animals from undergoing severe, and sometimes agonizing, suffering in laboratories. The U.S. National Cancer Institute replaced mice used to screen drugs for anti-cancer properties with cultured human cells, actually producing more accurate indications. Other government bodies also reduced their requirements for animal testing. The European Economic Community, the forerunner of the European Union, legislated to require that alternatives to animals be used whenever possible and set up the European Committee to Validate Alternative Models to determine when such alternatives should be used.[58] It looked as if the LD50, LC50, and Draize tests were on their way out.

In the 1990s the Organisation for Economic Co-operation and Development (OECD) began a program to standardize testing around the world. In the long run this will reduce the number of animals that need to be used because testing will not have to be repeated in each country to comply with minor variations in its regulations. At present, though, the OECD guidelines still require the use of animals, sometimes in large numbers. To register a pesticide, for example, requires many different studies, which take about three years and use, in total, 8,000 to 9,000 animals. There are an estimated 50,000 chemicals in use, 95 percent of which have yet to be fully evaluated. To evaluate them all by present methods would require 400 million animals.[59] The OECD is currently engaged in reducing the number of animals used in its tests and refining such tests so that they cause less suffering; but this is a long and slow process.

In the United States, the animal protection movement had been focused more on the use of animals in laboratories than on animals in factory farms, but in the 1990s that began to change, as organizations like People for the Ethical Treatment of Animals and the Humane Society of the United States challenged factory farming and the suffering it inflicted on animals. I supported that shift, because the number of animals raised and killed for food is hundreds of times greater than the number of animals used in laboratories. I began writing extensively about animals in factory farms and I no longer kept a close watch on what was happening to animals in laboratories. I assumed, though, that the trends begun in the 1990s toward phasing out the tests that had inflicted so much pain on animals had continued, and this seemed confirmed by some steps that were important enough to catch my attention. In 2002, the U.S. Environmental Protection Agency announced that it was suspending the use of conventional LD50 tests to test chemicals and was instead adopting more refined tests that, though still involving animals, would use smaller numbers and subject them to less suffering.[60] The European Union implemented a series of bans on the testing of cosmetics on animals, culminating in 2013 with a prohibition on the sale of cosmetics or ingredients of cosmetics that had been tested on animals after that date, no matter where in the world that testing took place. And in the United States in 2015, the National Institutes of Health ended its support for invasive research on chimpanzees. After having previously allowed nearly five hundred chimpanzees who had been held for biomedical research to live out the rest of their lives in sanctuaries, it finally did the same for a group of fifty it had been keeping in reserve for urgent requests—none of which were made.[61] Then in 2019 the Environmental Protection Agency announced that it would reduce studies that involve testing on mammals by 30 percent by 2025, and cease conducting or funding studies on mammals by 2035, unless they were authorized on a case-by-case basis. The EPA is the first U.S. federal agency to set a deadline for phasing out a major form of animal use.[62]

It was therefore with dismay that I discovered recently that LD50 tests have not disappeared. In vitro tests are now more commonly used to screen out substances that are toxic to cells, and the substances thus eliminated will no longer be tested on animals. That has reduced the number of animals used in LD50 tests. But substances that are not toxic to cells could still be toxic to entire systems, like the neurological system, so they are still being tested on groups of animals using some form of the LD50.[63] In the European Union in 2019 (at the time of writing, the most recent year for which figures are available), 31,654 animals were poisoned in LD50 and LC50 tests.[64] The United Kingdom has published figures for 2019, showing that the same tests were used on 10,558 animals.[65] As already noted, the United States does not keep figures on specific uses of animals, so the number used there is unknown.

We can also see the process of refinement and reduction at work with the Draize test. Here too, I was shocked to find that, forty years after Spira persuaded Revlon and Avon to put funds into finding an alternative, substances are still being put into the eyes of rabbits to test them for irritancy or corrosive effects. In a 2013 survey of the testing of products that might damage the eye, Leandro Texeira and Richard Dubielzig note that animal advocates have objected strongly to the Draize test and add that there has also been "major criticism" from toxicologists, who claim the Draize test is unreliable in predicting chronic toxic reactions, and its grading system has no objective scale of quantification. Texeira and Dubielzig list several ways of testing eye toxicity without using living animals, including the Eytex system. This method is, they say, quantitative and reproducible, and it correlates to a high degree with the Draize test. Nevertheless, "The Draize test is still the official model for eye irritation and toxicology studies worldwide."[66] It does seem, though, that recently more rapid progress is being made. European Union statistics show that the number of animals used in tests labeled "eye irritation/corrosion" fell from 4,208 in 2005 to 1,518 in 2015 and 474 in 2019.[67] Again, we have no information about the numbers of animals in the United States who had substances tested on their eyes.[68] In 2017, however, the

OECD adopted guidelines for eye-irritancy testing that included the use of pain relief when substances are tested on live animals. In 2021 these guidelines were further updated to minimize the number of animals used.[69]

As a result of all these developments in toxicity testing, an immense amount of needless pain and suffering has been avoided. Precisely how much is hard to say, but millions of animals would have suffered each year in tests that will now not be performed. That is a significant achievement for which the animal movement can take much credit; but there is still more work to be done before these cruel tests are ended worldwide. Stopping this waste of animal lives and animal pain should not be difficult if people really want to do it. Developing completely adequate alternatives to all tests for toxicity using animals will take longer, but it should be possible. Meanwhile there is a simple way to cut down the amount of suffering involved in such tests: Until we have developed satisfactory alternatives, as a first step, we could do without any new but potentially hazardous substances that are not essential to our lives.

MEDICAL EXPERIMENTS

WHEN EXPERIMENTS ON ANIMALS FALL UNDER THE HEADING "medical" we are inclined to think that any suffering they involve must be justifiable because the research is contributing to prolonging human lives and alleviating suffering. But we have already seen that pharmaceutical companies producing new therapeutic drugs are likely to be motivated by the desire for maximum profit rather than maximum good to all, and that researchers, too, tend to continue to devise variations on experiments, which may have more to do with achieving success in their career, by obtaining more research grants and publishing more articles, than with improving human health.

Consider, as yet another example of the way in which experiments come to be carried out more for the benefit of the researchers than to benefit human health, experiments relating to the experimental induction of shock in animals (meaning, this time, not electric shock but the mental and physical state of shock that often occurs after a severe injury). As long ago as 1946 a researcher in the field, Magnus Gregersen of Columbia University, surveyed the literature and found over eight hundred published papers dealing with experimental studies of shock. He describes the methods used to induce shock:

> The use of a tourniquet on one or more extremities, crush, compression, muscle trauma by contusion with light hammer blows, Noble-Collip drum [a device in which animals are placed and the drum rotated; the animals tumble repeatedly to the bottom of the drum and are injured], gunshot wounds, strangulation or intestinal loops, freezing, and burns.

Gregersen also notes that hemorrhaging has been "widely employed" and that "an increasing number of these studies has been done without the complicating factor of anesthesia." He is not, however, pleased by all this diversity and complains that the variety of methods makes it "exceedingly difficult" to evaluate the results of different researchers; there is, he says, a "crying need" for standardized procedures that will invariably produce a state of shock.[70]

Jump forward to 1974, and experimenters were still trying to determine what injuries might be inflicted to produce a satisfactory "standard" state of shock that will enable them to produce a suitable animal model. After nearly three decades of experiments designed to produce shock in dogs by causing them to hemorrhage, more recent studies indicated that hemorrhage-induced shock in dogs is not like shock in humans. Noting these studies, researchers at the University of Rochester caused hemorrhaging in pigs, which they thought might be more like humans than dogs are in this respect. That required them to determine what volume of blood loss might be suitable for the production of experimental shock in pigs.[71]

Now let's go forward another four decades to 2013, almost seventy years after Gregersen lamented the lack of a standardized procedure for producing shock. A. Fülöp and a team of colleagues from Semmelweis University, in Budapest, reviewed animal models of hemorrhagic shock and found that:

> Although a wide spectrum of species and experimental models are available to researchers, it is rather difficult to create an ideal animal model to study hemorrhagic shock. A major challenge for investigators is the generation of a system which is simple, easily reproducible and standardized, while being an accurate replica of the clinical situation. . . . The different genetic background of different species can result in distinct systemic responses to the same insult. Therefore, it is by no means certain that the results can be applied to human cases.[72]

Addiction is another area in which there has been a longstanding search for an animal model, going back to studies on alcoholism following the repeal of prohibition in the United States, with surges of research with each new form of addiction. Here is one example, from the 1980s:

At the University of Kentucky, beagles were used to observe withdrawal symptoms from Valium and Lorazepam, a similar tranquillizer. The dogs were forced to become addicted to the drug and then, every two weeks, the tranquillizers were withdrawn. Withdrawal symptoms included twitches, jerks, gross body tremors, running fits, rapid weight loss, fear, and cowering. After forty hours of Valium withdrawal, "numerous tonic-clonic convulsions were seen in seven of nine dogs. . . . Two dogs had repeated episodes of clonic seizures involving the whole body." Four of the dogs died—two while convulsing and two after rapid weight loss. Lorazepam produced similar symptoms but not convulsive deaths. The experimenters reviewed experiments going back to 1931 in which barbiturate and tranquillizer withdrawal symptoms had been observed in rats, cats, dogs, and primates.[73]

In an even more bizarre example of drug research, also in the 1980s, Ronald Siegel at the University of California at Los Angeles chained two elephants to a barn. The female elephant was used in range-finding tests "to determine procedures and dosages for LSD administration." She was given the drug orally and by dart gun. After this the experimenters dosed both elephants every day for two months and observed their behavior. High doses of the hallucinogen caused the female to fall down on her side, trembling and barely breathing, for one hour. The high doses caused the bull elephant to become aggressive and charge Siegel, who described such repeated aggressive behavior as "inappropriate."[74]

One episode in this grim tale of drug experimentation does, at least, demonstrate the efficacy of protest. In a paper published in 1976, researchers at Cornell University Medical College described feeding large doses of barbiturates to cats by means of tubes surgically implanted in their stomachs. They then abruptly stopped the barbiturates. Here is their description of the withdrawal symptoms:

> Some were unable to stand. The "spread eagle posture" was seen in animals displaying the most severe abstinence signs and the most frequent grand mal type convulsions. Almost all of these animals died during or soon after periods of continuous convulsive activity. . . . Rapid or labored respiration was often noted when other abstinence signs were most intense. . . . Hypothermia was noted when animals were weakest, especially after persistent seizures and when near death.[75]

Although barbiturate abuse had been a serious problem a few years earlier, by the mid-1970s the use of barbiturates was severely restricted, and abuse had declined. It continued to decline during the fourteen years that these cat experiments at Cornell continued. In 1988, after animal rights campaigners picketed the laboratory at which the studies were being conducted and wrote letters to the funding agencies, the press, the university, and legislators, Cornell wrote to the funding body, the National Institute on Drug Abuse, to say that the university would forfeit a new $530,000 research grant that would have paid for three more years of experiments.[76]

Alcohol research is another major area of addiction studies. The initial problem is that most animals other than humans don't voluntarily become addicted to alcohol. Various techniques are therefore employed to make them addicted, including feeding them an alcohol-laced all-liquid diet so that they have to consume alcohol or starve; forcing them to inhale alcohol vapor; and gavage, which, as we have already seen, is a distressing and often painful experience that sometimes causes fatal injuries. After forcing animals to become addicted to alcohol, researchers then cut it off, to study withdrawal symptoms and, in some cases, add stress to the withdrawal.

Kathryn Harper and other researchers at the University of North Carolina at Chapel Hill forced both adolescent and adult rats to consume alcohol as part of a liquid diet—a diet that, the authors say in a paper published in 2017, had been used by their lab "for several decades to establish the effects of alcohol withdrawal on anxiety-like behavior." When the alcohol was finally withdrawn, "half the rats were exposed to 1-h restraint stress in plastic rat decapicones." (Decapicones are soft plastic cones with a hole at the end—similar to cake icing funnels. The name comes from the fact that they are normally used to hold rodents in place before beheading—that is, decapitating—them. Naturally, keeping rats or mice in these cones for long periods causes stress.) The researchers measured the stress levels of the rats and found that while the addition of stress during withdrawal increased anxiety in adult rats, in adolescent rats it had the opposite effect, returning their behavior to that of rats who had not been addicted to alcohol. But, the researchers noted, "These behavioral effects were moderate in magnitude, and additional studies to replicate these findings and to identify optimal conditions are warranted."[77]

Are these additional studies warranted? Do we need to identify optimal conditions for observing the differences in the behavior and the amygdala responses of adolescent and adult rats who have been addicted to alcohol and then had it cut off, and have then been put in a soft plastic cone, unable to move, for an hour? Why, exactly, is that worth the time and efforts of highly trained and intelligent scientists? Why is that the best possible use of funds from the National Institute on Alcohol Abuse and Alcoholism? And why does it justify inflicting suffering on animals?

In a trenchant critique of the use of animal models of addiction, Matt Field and Inge Kersbergen of the department of psychology and the School of Health and Related Research at the University of Sheffield argue that the contribution of animal models of addiction in identifying and developing treatments that benefit humans has been "consistently misrepresented and oversold," which has resulted in "a huge waste of resources." They suggest that addiction may be a uniquely human phenomenon, dependent on language, and also affected by environmental and social networks that cannot be modeled in animals. Perhaps even more serious, they claim, is the way in which the emphasis on animal models of addiction has "misled us about the very nature of addiction in humans."[78] Unless these claims can be decisively refuted, it is reasonable to believe that almost a century of producing animal models of addiction has not only inflicted pain and suffering on animals, and been a waste of resources, but has also done addicted humans more harm than good.

CONDITIONED ETHICAL BLINDNESS

HOW CAN THESE THINGS HAPPEN? HOW CAN PEOPLE WHO ARE not sadists spend their working days terrifying infant monkeys, giving mice inescapable electric shocks, and addicting cats to drugs or alcohol? How can they then remove their white coats, wash their hands, and go home to dine comfortably with their families? How can taxpayers allow their money to be used to support these experiments?

The answer, I believe, to these questions lies in the unquestioned acceptance of speciesism. We tolerate cruelties inflicted on members of other species that would outrage us if performed on members of our own species. Speciesism allows researchers to regard the animals they experiment on as items of equipment, laboratory tools rather than living, suffering creatures.

In addition to the general attitude of speciesism that experimenters share with other citizens, some special factors also help to make possible the experiments I have described. Foremost among these is the immense respect that people still have for scientists. Although most thinking people now realize that science and technology have a dark side, many people are still in awe of anyone who wears a white coat and has a Ph.D. In a famous experiment, Stanley Milgram demonstrated that ordinary people will obey the directions of a white-coated researcher to administer what appears to be electric shock to a human subject as "punishment" for failing to answer questions correctly, and they will continue to do this even when the human subject cries out and pretends to be in great pain. A recent replication of this experiment found a similar result.[79] If this can happen when the participants believe they are inflicting pain on a human being, how much easier is it for students to push aside their initial qualms when their professors instruct them to perform experiments on animals?

Donald Barnes was for several years a principal investigator in the U.S. Air Force School of Aerospace Medicine. At Brooks Air Force Base, in San Antonio, Texas, he was in charge of training monkeys to operate something called a primate equilibrium platform. The monkeys are locked into a chair that is part of the platform, which can be made to pitch and roll like an airplane. They are then trained to keep the platform horizontal, which they can do by operating a control stick. This training is done in stages, during which the monkeys receive electric shocks until eventually they learn what they are supposed to do. Over the forty days of training, each monkey receives thousands of electric shocks. Once the monkeys have been trained, they are exposed to radiation and chemical warfare agents and evaluated on their ability to keep the platform level. The idea is to see whether pilots exposed to radiation and toxic chemicals would be able to continue to complete their mission. (If this sounds familiar, it may be because you have seen *Project X*, a 1987 movie starring Matthew Broderick and Helen Hunt, which was loosely based on the experiments Barnes conducted.)

Here is an example of just one of the experiments carried out on monkeys who had been trained to use the primate equilibrium platform, taken from a 1987 United States Air Force School of Aerospace Medicine report titled "Primate Equilibrium Performance Following Soman Exposure: Effects of Repeated Daily Exposures to Low Soman Doses."[80] Soman is another name for nerve gas, a chemical warfare agent that caused terrible agony to troops in the First World War but fortunately has been very little used in warfare since then. The report begins by referring to several previous reports in which the same team of investigators studied the effects of "acute exposure to soman" on performance in the primate equilibrium platform. This particular study, however, is on the effect of low doses received over several days. The monkeys in this experiment had been operating the platform "at least weekly" for a minimum of two years and had received various drugs and low doses of soman before, but not within the previous six weeks. The experimenters calculated the doses of soman that would be sufficient to reduce the monkeys' ability to operate the platform. For the calculation to be made, of course, the monkeys would have been receiving electric shocks because of their inability to keep the platform level. Although the report is mostly concerned with the effect of the nerve poison on the performance level of the monkeys, it does give some insight into other effects of chemical weapons:

> The subject was completely incapacitated on the day following the last exposure, displaying neurological symptoms including gross incoordination, weakness, and intention tremor. . . . These symptoms persisted for several days, during which the animal remained unable to perform the PEP task.[81]

Barnes estimates that he irradiated about one thousand trained monkeys during his 14 years in this position. Then, as he later wrote, something changed:

> For some years, I had entertained suspicions about the utility of the data we were gathering. I made a few token attempts to ascertain both the destination and the purpose of the technical reports we published but now acknowledge my eagerness to accept assurances from those in command that we were, in fact, providing a real service to the U.S. Air Force and, hence, to the defense of the free world. I used those assurances as blinkers to avoid the reality of what I saw in the field, and even though I did not always wear them comfortably, they did serve to protect me from the insecurities associated with the potential loss of status and income. . . . And then, one day, the blinkers slipped off, and I found myself in a very serious confrontation with Dr. Roy DeHart, Commander, U.S. Air Force School of Aerospace Medicine. I tried to point out that, given a nuclear confrontation, it is highly unlikely that operational commanders will go to charts and figures based upon data from the rhesus monkey to gain estimates of probable force strength or second strike capability. Dr. DeHart insisted that the data will be in-

valuable, asserting, "They don't know the data are based on animal studies."[82]

Barnes, whose training was in psychology, later compared what happened to him during his training with what happens when a rat is conditioned to press a lever to get a food reward:

> I represented a classic example of what I choose to call "conditioned ethical blindness." My entire life had consisted of being rewarded for using animals, treating them as sources of human improvement or amusement. . . . During my sixteen years in the laboratory the morality and ethics of using laboratory animals were never broached in either formal or informal meetings prior to my raising the issues during the waning days of my tenure as a vivisector.[83]

Steven Pinker, a professor of psychology at Harvard University, tells a similar story of ethical blindness when, as a student, he got a summer job as a research assistant in an animal behavior lab. One evening the professor in charge told him to try a new experiment on a rat. Pinker was instructed to put the rat in a box with an electrified floor and a timer that would shock the rat every six seconds unless it pressed a lever, which would halt the shocks for ten seconds. Rats, he was told, catch on quickly, and the rat would soon learn to press the lever in time to avoid receiving shocks. So all Pinker had to do was put the rat in the box, start the timer, and go home. But when Pinker arrived the next morning, the rat had "a grotesque crook in its spine and was shivering uncontrollably," and within a few seconds, it jumped up, having received a shock. Pinker realized that the rat had not learned to press the lever and had spent the whole night being shocked every six seconds. He took it out of the box and brought it down to the laboratory veterinarian, but the rat died. In Pinker's own words: "I had tortured an animal to death." Pinker describes this as "the worst thing I have ever done" and also says that as the experiment was being explained to him, he already sensed that it was wrong because even if the rat had learned to press the lever, it would have spent twelve hours in constant anxiety. But, he confesses: "I carried out the procedure anyway, reassured by the ethically spurious but psychologically reassuring principle that it was standard practice."[84]

It is not only the experimenters themselves who suffer from conditioned ethical blindness. Research institutions sometimes answer critics by telling them that they employ a veterinarian to look after the lab animals. Such statements are supposed to provide reassurance because of the widespread belief that veterinarians are people who care about animals and it would be contrary to the ethics of their profession to let them suffer unnecessarily. Regrettably, this is not always so. No doubt many veterinarians do go into the field because they care about animals, but it is difficult for people who really care about animals to go through a course of study in veterinary medicine without having their sensitivity to animal suffering blunted. Those who care most may not be able to complete their studies. One former veterinary student wrote to an animal welfare organization:

> My life-long dream and ambition to become a veterinarian dissipated following several traumatic experiences involving standard experimental procedures utilized by the dispassionate instructors of the Pre-Vet school at my state university. They felt it was perfectly acceptable to experiment with and then terminate the lives of all the animals

> they utilized, which I found revoltingly unacceptable to my own moral code. After numerous confrontations with these heartless vivisectionists, I painfully decided to pursue a different career.[85]

In the United States in late 1965 and early 1966, articles in *Life* and *Sports Illustrated*, then both very widely read magazines, described how family dogs had been stolen from homes and sold to laboratories. There was a huge public response, which led to moves being made to pass legislation to protect laboratory animals.[86] The American Veterinary Medical Association testified to congressional committees that while it favored legislation to stop the stealing of pets for subsequent sale to laboratories, it was opposed to the licensing and regulation of research facilities, since this could interfere with research. The basic attitude of the profession was, as an article in the *Journal of the American Veterinary Medical Association* put it, that "the raison d'etre of the veterinary profession is the over-all well-being of man—not lower animals."[87] Given that ideology, we should not be surprised that veterinarians were part of the experimental teams that performed many of the experiments described in this chapter. The report from the primate equilibrium platform experiment, for example, states: "Routine care of the animals was provided by the Veterinary Sciences Division, USAF School of Aerospace Medicine."

Once a pattern of animal experimentation becomes the accepted mode of research in a particular field, the process is self-perpetuating. In fields dominated by professors who have built their careers on experimenting on animals, not only publications and promotions but also the awards and grants that finance research become geared toward animal experiments. A proposal for a new experiment with animals is something that those reviewing applications for research grants will be ready to support. New methods that do not make use of animals will be less familiar and less likely to receive support.

Not all scientists are ethically blind, of course. A few are led by unexpected outcomes of their research to think afresh about the animal subjects of their work. Ecologists studying the behavior of wild animals have taken advantage of developments in technology like GPS tracking devices, and these devices can now be made sufficiently small and lightweight for attachment to birds. Dominique Potvin, a Senior Lecturer in ecology at the University of the Sunshine Coast, in Queensland, was working with a team to develop a lightweight harness with a miniature tracking device that could be attached to the Australian magpies they were studying. They then ran a pilot study to test the devices in the field. Over a period of ten weeks, they accustomed a group of magpies to visiting a feeding station, where they were then able to trap five of them, fit them with the harness and tracking device, and release them. The released birds began pecking at their own harnesses but were unable to remove them. A young magpie without a harness then approached another young magpie with a harness and pecked at its harness. This was ineffective, but after a few minutes an adult female approached and began pecking at the harness on the young magpie. After about ten minutes, the harness came off. The same thing happened with two other birds. The next day the remaining two magpies on whom harnesses had been fitted were seen still wearing them, but by the third day, all the harnesses had been removed.

The striking result of this study was not information gained from the tracking devices about the birds' movements, but the cooperative and altruistic behavior of the magpies who had not been fitted with harnesses but helped other magpies to remove their harnesses. This involved, as Potvin and her colleagues said, "complex cognitive problem solving." In conclusion, they mentioned the need to consider "the implications of these responses to ethical ecological study."[88]

After reading about this study, I wrote to Potvin to ask whether, in view of the magpies' clear refusal of consent to participate in the study, she was planning another attempt to fit tracking devices that the magpies could not remove. She replied: "We never seem to know how animals might respond to trackers, which is why we pilot on a handful of individuals first. Obviously they were not fans, so it would not be ethical (nor scientifically reasonable) to continue to try and track them with similar devices."[89]

NOT EVEN GOOD SCIENCE

WHEN WE READ REPORTS OF EXPERIMENTS THAT CAUSE PAIN AND are apparently not even intended to produce results of real significance, we are at first inclined to think that there must be more to what is being done than we can understand—that the scientists must have some better reason for what they are doing than their publications indicate. When we go more deeply into the subject, however, we find that what appears trivial on the surface very often really is trivial. Occasionally, experimenters lower their guard and admit it. Harry Harlow, whose experiments on maternal deprivation in monkeys we encountered at the beginning of this chapter, was for twelve years the editor of the *Journal of Comparative and Physiological Psychology*, a journal that has published many reports of painful experiments on animals. At the end of this period, during which Harlow estimated he reviewed about 2,500 manuscripts submitted for publication, he wrote, in a semi-humorous farewell note, that "most experiments are not worth doing and the data attained are not worth publishing."[90]

Harlow wrote that in 1962. Since then serious research has

been done into how much biomedical research—both in general and involving animals—is worth doing. The answers should alarm everyone, even if they are not concerned about reducing animal suffering. There is now general acknowledgment that the biomedical and behavioral sciences face a "replication crisis": If scientific results cannot be replicated, they do not add to our knowledge. Yet studies attempting to replicate findings in biomedicine and psychology have found that from 50 to 90 percent are not replicable.[91] In a scathing review of research in their own field, C. Glenn Begley and John P. A. Ioannides, both eminent medical scientists, accepted that 85 percent of biomedical research is wasted. They analyzed over 4,445 data sets involving animal studies of neurological disease and found "there were just too many positive results published to be true." Begley and Ioannides offered the same underlying explanation as the team led by van der Naald at Utrecht University, which showed that published reports accounted for only 26 percent of all animals used in research: a bias in favor of publishing positive results. Begley and Ioannides urged that funding agencies need to be much more stringent in the research that they approve.[92] Another study has found that the United States alone is spending $28 billion each year on preclinical research that cannot be reproduced.[93]

If that is true of biomedical research in general, then an even smaller proportion of research with animals will benefit humans, because what works with animals often will not work with humans. Richard Klausner, former director of the U.S. National Cancer Institute, said, "We have cured mice of cancer for decades—and it simply didn't work in humans."[94] Nor are the failures to translate findings on animals to humans specific to cancer. A collaborative research team comprising scientists from leading U.S. and Canadian institutions investigated the use of mice to find treatments for inflammatory disorders, which can be fatal for people who have been severely injured or burned, or who have an infection that has led to sepsis. The team examined the outcomes of nearly 150 clinical trials that tested what seemed, from experiments with mice, to be promising ways of blocking the inflammatory response in critically ill human patients. Their verdict: "Every one of these trials failed." The team concluded that their study "supports higher priority to focus on the more complex human conditions rather than relying on mouse models to study human inflammatory diseases."[95]

Frances Cheng, a researcher who earned her doctorate doing experiments on animals at Case Western Reserve University, a leading U.S. center of biomedical research, has described how difficult it is for junior researchers to push back against the paradigm of experimenting on animals. The team she was working with had received a grant to carry out experiments that involved inducing heart failure in rats in order to study the role played by a high-fat diet on the function of the heart, believing that this could improve our understanding of the links between diet and heart disease in humans. Only after the experiment had been carried out and submitted to a journal did Cheng discover that rats metabolize fat differently from humans, and so her findings would not apply to humans—something that the reviewers of the team's grant application had failed to notice.

The view of Cheng's advisors was that the paper she had written was simply reporting the findings of their research on rats and did not claim that these findings had any value for improving human health. But Cheng was dismayed to realize that she had, as she put it, "published a misleading study and harmed animals in useless research." She talked to her Ph.D. committee about what had happened and was surprised to find that it didn't trouble them. "Someone will find your work useful somehow, someday," one of them said. Another told her: "Your job as a Ph.D.

student is to graduate, it's not your job to think about this." A third asked: "What else can we do [if we don't use animals]?" Cheng realized she wasn't getting any real answers as to how to overcome the problem of animal data not translating to humans, and she realized that none of the senior scientists knew the answer either. "They've just been trained in using animals for so long," she said, "that either they kid themselves into believing it or they just hold on to this last hope that it will help humans, somehow, someday."

Elias Zerhouni, a former director of the U.S. National Institutes of Health, said, regarding the assumption that using animal models was the best way to learn how to prevent or cure human diseases: "We all drank the Kool-Aid on that one, me included." He added: "It hasn't worked, and it's time we stopped dancing around the problem. . . . We need to refocus and adapt new methodologies for use in humans to understand disease biology in humans."[96] The tragedy is that Zerhouni admitted his mistake in 2013, when he no longer had the power to change the direction of NIH. Meanwhile the animal research industry powers on, causing immense suffering to animals and wasting tens of billions of dollars in research funds that might, if used more effectively, have cured some of the diseases that afflict us.

In the United Kingdom, there are signs that the biomedical research community is reconsidering the central place of animals. A joint report by the UK Bioindustry Association and the Medicines Discovery Catapult (a government-funded organization that seeks to stimulate innovation in biomedical research) has concluded that pre-clinical research has been "patient free, and relied on animal modes of disease and toxicology that were a poor approximation of humans." If the UK bioindustry is to compete with China and India, the report says, it must "humanize" drug discovery by starting with patients and creating candidate drugs that "are highly selective for proven human disease targets in well-defined patient sub-groups, not animal targets." The report recommends the development of human in vitro models to provide evidence of toxicity and efficacy in humans, not animals.[97]

INEFFECTIVE REGULATION

IF REVELATIONS ABOUT THE WASTE OF TENS OF BILLIONS OF DOLlars have not succeeded in eliminating most animal research, one might think that laws protecting animals would prevent at least the worst of the suffering I have described in this chapter. Laws vary among countries, of course, but in the United States, animal experimenters, working together with the major industries that produce and use animals for research, have been effective in preventing standard anti-cruelty legislation from applying to animals used in research. David Baltimore, who was president of the California Institute of Technology until 2006, once told a national meeting of the American Association for the Advancement of Science about the "long hours" that he and his colleagues had spent fighting regulation of their research.[98] The basis for his opposition to such regulation was made clear some years earlier, when he appeared on a television program with the late Harvard philosopher Robert Nozick and other scientists. Nozick asked whether the fact that an experiment would kill hundreds of animals is ever regarded, by scientists, as a reason for not performing it. One of the scientists answered: "Not that I know of." Nozick pressed his question: "Don't the animals count at all?" A scientist countered: "Why should they?" At this point Baltimore interjected that he did not think that experimenting on animals raised a moral issue at all.[99] Unfortunately, brilliance in science is no guarantee of a sound grasp of ethics. Today, at least in the English-speaking philosophical world, it would be difficult to find professional philosophers who would agree with Baltimore's statement.

American scientists have, so far, been extraordinarily intransigent about public oversight of what they do to animals. They have been successful in squelching even minimal regulations to protect animals from suffering in experiments. In the United States, the only federal law that could effectively protect animals used in research is the Animal Welfare Act, passed in 1966 after the previously mentioned uproar over the revelations that companion animals were being stolen and sold to laboratories. That law sets the standards for the transportation, housing, and handling of animals sold as pets, exhibited, or intended for use in research. So far as actual experimentation is concerned, however, it allows the researchers to do exactly as they please. This is quite deliberate: When the act was passed the U.S. Congress conference committee explained its purpose was

> to provide protection for the researcher in this matter by exempting from regulation all animals during actual research or experimentation. . . . It is not the intention of the committee to interfere in any way with research or experimentation.[100]

One section of the law requires that those private businesses and other organizations that register under the act must file a report stating that when painful experiments are performed without the use of pain-relieving drugs, this is necessary to achieve the objectives of the research project. No attempt is made, however, to assess whether these "objectives" are sufficiently important to justify inflicting pain. Under these circumstances the requirement does no more than make additional paperwork—a major complaint among experimenters. They can't, of course, give animals the continuous electric shocks that will produce a state of helplessness if they anesthetize them at the same time; nor can they produce depression in monkeys while keeping them happy or oblivious with drugs. So in such cases the experimenters can truthfully state that the objectives of the experiment cannot be achieved if pain-relieving drugs are used. They can then go on with the experiment as they would have done before the act came into existence.

So we should not be surprised that, for instance, the report of the primate equilibrium platform experiment with soman (see p. 75, above) should be prefaced with the following statement:

> The animals involved in this study were procured, maintained, and used in accordance with the Animal Welfare Act and the "Guide for the Care and Use of Laboratory Animals" prepared by the Institute of Laboratory Animal Resources–National Research Council.

Similarly, the published account of the experiment conducted in China by Teng and his colleagues, which attempted to cause depression in monkeys by isolating them in cages for seventy days, includes this statement:

> Animals were maintained under an experiment protocol approved by the Ethics Committee of Chongqing Medical University (approval no.: 20180705) in accordance with the recommendations of "The use of nonhuman primates in research" and "Guide for the Care and Use of Laboratory Animals."[101]

Perhaps, though, it is the final words of the "Ethics Declaration" in the publication by Yin's team at the Beijing Institute of Pharmacology and Toxicology (p. 52 above), describing how they isolated and stressed monkeys in order to induce depression in them, that best shows how misleading such declarations can be:

> All procedures were in strict accordance with the guidelines of the National Institutes of Health Guide for the Care and Use of Laboratory Animals (NIH Publications No. 80–23, revised in 1996) and the animal study was reviewed and approved by the National Animal Research Authority (China) and Beijing Institute of Pharmacology and Toxicology. All efforts were made to reduce the number of monkeys used and to minimize animal suffering.[102]

Given that the whole point of the experiment was to make the animals suffer severely enough to become a model of depression, we can easily see that such statements are no guide to how much suffering the experiment inflicted on the animals.

The complete absence of effective regulation in the United States is in sharp contrast to the situation in many other developed nations. In Britain, for example, no experiment can be conducted without a license granted by the Secretary of State for Home Affairs, and the Animals (Scientific Procedures) Act, 1986, expressly directs that in determining whether to grant a license for an experimental project, "the Secretary of State shall weigh the likely adverse effects on the animals concerned against the benefit likely to accrue." In Australia, the Code of Practice developed by the leading governmental scientific bodies (equivalent to the National Institutes of Health in the United States) requires that all experiments must be approved by an Animal Experimentation Ethics Committee. These committees must include a person with an interest in animal welfare who is not employed by the institution conducting the experiment, and an additional independent person not involved in animal experimentation. The committee must apply a detailed set of principles and conditions that include an instruction to weigh the scientific or educational value of the experiment against the potential effects on the welfare of animals. In addition, anesthesia must be used if the experiment "may cause pain of a kind and degree for which anesthesia would normally be used in medical or veterinary practice." The Australian Code of Practice was originally applicable only to researchers obtaining government grants, but it has now been incorporated into the law of each state and territory in Australia and so is legally binding on anyone carrying out experiments on animals.[103] Sweden also requires experiments to be approved by committees that include lay members. In 1986 the U.S. Congress's Office of Technology Assessment surveyed laws in Australia, Canada, Japan, Denmark, Germany, the Netherlands, Norway, Sweden, Switzerland, and the United Kingdom, and concluded:

> Most of the countries examined for this assessment have laws far more protective of experimental animals than those in the United States. Despite these protections, animal welfare advocates have been applying considerable pressure for even stronger laws, and many countries, including Australia, Switzerland, West Germany, and the United Kingdom, are considering major changes.[104]

Stronger laws have in fact already been passed in Australia, the United Kingdom, and the European Union since that statement was made.

I hope this comparison will not be misunderstood. It is not intended to show that all is well with animal experimentation in the United Kingdom, or Australia, or the European Union. That would be far from the truth. In those countries the "balancing" of potential benefits against harm to the animals is still carried out with the assumption that human interests always count much more than the interests of animals. I have compared the situation in the United States with that in other countries only in order to show that American standards in this matter are abysmal, not just by the standards of animal liberationists but by those accepted by the scientific communities of other major developed nations. This is, I believe, because the United States political system is more open to distortion by lobbyists—in this case from animal researchers, companies that profit from animal research, and also, shamefully, the American Veterinary Medical Association—than other democracies.

The history of the failure to regulate animal experimentation in the United States is instructive. In 1970, the Animal Welfare Act was amended to include all warm-blooded animals used in experiments. Yet the Secretary of Agriculture, in charge of the department that administers the Animal Welfare Act, never issued regulations to cover rats, mice, or birds, which constitute at least 95 percent of all warm-blooded animals used in U.S. laboratories. Nor was the department's inspection service ever allocated sufficient funds to inspect them, so they were ignored. As the Office of Technology Assessment (OTA) reported in 1986: "Funds and personnel for enforcement have never lived up to the expectations of those who believe the primary mission of the existing law to be the prevention or alleviation of experimental animal suffering."[105] OTA staff checked one list of 112 testing facilities and found that 39 percent were not even registered with the branch of the Department of Agriculture that inspects laboratories. Moreover, the OTA report stated that this is probably a conservative estimate of the real number of unregistered, and hence totally uninspected and uncontrolled, animal laboratories.[106]

In 1998 the American Anti-Vivisection Society took the secretary of agriculture to court, claiming that there was no legal basis for not covering rats, mice, and birds used in laboratories. The Society won, and the Department of Agriculture prepared new regulations to cover these animals—but before they could come into effect, lobbyists for the research industry succeeded in getting Congress to pass a new amendment that said, astonishingly, that for the purposes of the Animal Welfare Act, the term "animal" excludes "birds, rats (of the genus *Rattus*), and mice (of the genus *Mus*) bred for use in research." That is still, at the time of writing, the situation in the United States. The failure to cover the animals most commonly used in experiments stands as an indictment of the obstructionist attitude of U.S. scientists and the U.S. animal experimentation industry to the most elementary improvements in the conditions of the vast majority of the animals they use.[107]

In 1985, the United States Animal Welfare Act was amended to set up Institutional Animal Care and Use Committees, or IACUCs. Stephen Pinker—who, as we saw earlier in this chapter, carried out, on the instructions of a professor, an experiment that he later regarded as torturing a rat to death—has stated that the introduction of these committees has transformed attitudes toward animals among scientists, to such an extent that "a scientist who was indifferent to the welfare of laboratory animals would be treated by his or her peers with contempt."[108] I wish it were so; but Pinker's claim that attitudes toward animals in American universities have changed so thoroughly since his student days is difficult to reconcile with several experiments described in this chapter, carried out by American researchers long after IACUCs were required. We could start with the experiments on maternal deprivation by Suomi, which continued until 2015; the work done by Donovan, Liu, and Wang stressing prairie voles; Shively and Neigh's involvement with the research team at Chongqing Medical University producing anxiety and depressive behavior in monkeys; and Harper and colleagues' experiments on rats and alcohol.

Emily Trunnell, whose assessment of research that attempted to create monkey models of depression I quoted earlier in the chapter, offers a different view of the attitude that prevails in American university departments that use animals in research. She started graduate school at the University of Georgia in 2012, when IACUCs had been part of American universities for more than twenty-five years. As the first in her family to go to a university, she was thrilled to be admitted to graduate school, and especially in neuroscience, her chosen field. She had no specific idea about what kind of research she would do. Her advisor suggested she do experiments on animals, since she would have a greater chance of getting her work published in journals than if she did in vitro work. She took that advice. At first she was excited to be working "with" animals, and she enjoyed spending time with the mice and rats, especially the rats. In an interview for a short documentary called "Test Subjects," Trunnell later said: "Going in every day to the lab, I just felt really important. . . . I thought, Okay, these are important experiments, and we use a few animals here and there, but it's for the good of science." After a time, however, she became uncomfortable with what she was doing, especially when one of her final experiments involved injecting rats with substances that, she believed, caused them considerable abdominal pain. It became apparent to her that she wasn't doing this for a greater good, to cure diseases: "The point of me designing this experiment, which would use animals, was to get my degree, and nothing else. To finish, so that I could be *Dr. Emily.*" She didn't think that was worth taking the lives of about 200 animals, which she had done. When she came to write up her dissertation, her task was to show how the experiments she'd done had benefited humans. Although, unlike Pinker, she had not been expressly instructed to cause pain to an animal, she became troubled by the ease with which she had been allowed to design and carry out experiments that did cause pain to rats and took their lives, despite providing very little justification in terms of the good the experiments might do. As she said in the film:

> I felt kind of embarrassed to talk about what I did. I have this feeling that somehow I should have known better, even though at the time I really had no way of knowing better, everybody in charge was doing all of this too. The message that you get is that if you even start to question—whether or not the experiment that I am doing, is that right, in the bigger sense?—that's very taboo. You can't question. Any science is good science. We're going to learn something, even if it's poorly designed, we're going to learn something from it. And to me, when someone's suffering, because of that, you need to take a step back.

When Trunnell completed her dissertation, she included a dedication:

> I dedicate this body of work to those who gave the ultimate sacrifice. Over 200 animals, mice and rats, were euthanized in the direct collection of primary research presented in the following chapters.

Subsequently she decided to change careers and became an advocate for animals. As a result, her major advisor refused to walk her across the stage to receive her degree at her graduation ceremony. "He said that he felt betrayed," she recalled, "and it would be awkward for him." She didn't want to make anyone uncomfortable, so she ended up not going to the ceremony.[109]

WHEN ARE EXPERIMENTS ON ANIMALS JUSTIFIABLE?

UPON LEARNING OF THE NATURE OF MANY OF THE EXPERIMENTS carried out, some people react by saying that all experiments on animals must be banned. Putting morality in such black-and-white terms is tempting because it eliminates gray areas and the need to say where the line is to be drawn. But it will not do to say "Never!" If we make our demands as absolute as this, the experimenters have a ready reply: Would we be prepared to let thousands of humans die if they could be saved by a single experiment on one mouse? The way to reply to this purely hypothetical question is to pose another: Suppose that the only way to save thousands of lives is to experiment on a human being; and, as it happens, a human infant has just been rescued from a car crash in which the infant's parents were killed. The infant has suffered major brain damage, however, and tests reveal that their cognitive abilities and capacity to suffer are inferior to those of a mouse, and there is no chance of any recovery. If the defender of experimenting on animals would not be prepared to use a human infant under those circumstances, then their readiness to use nonhuman animals reveals an unjustifiable form of discrimination on the basis of species. (I have specified that the parents have died in the crash to avoid the objection that the parents would suffer if the child were used in an experiment. Specifying the case in this way is, if anything, overgenerous to those defending the use of nonhuman animals in experiments, since mammals intended for experimental use are usually separated from their mothers at an early age, when the separation causes distress for both mother and child.)

To avoid misunderstanding: This is a hypothetical example, of course. It is not proposed as an alternative to animal experiments. Instead it is a means of testing whether the defense of animal experimentation is based on capacities that are unique to human beings, or simply on membership of our species. But it is not a far-fetched example. There are, unfortunately, many intellectually disabled human beings, some of them long since abandoned by their parents and other relatives and, sadly, sometimes unloved by anyone else. The anatomy and physiology of these people are in nearly all respects identical to those of normal humans. If, therefore, we were to force-feed them large quantities of floor polish or drip concentrated solutions of household products into their eyes, we would have a much more reliable indication of the safety of these products for humans than we now get by attempting to extrapolate the results of tests on a variety of other species. The LD50 tests, the Draize eye tests, and others described earlier in this chapter could have told us more about how humans would react to the substances being tested if they had been carried out on severely brain-damaged humans instead of dogs, rabbits, or mice.

Occasionally medical experiments performed on human beings incapable of giving their consent come to light. When, for example, it was revealed that, as part of a series of medical experiments commenced in 1956 and ended only in 1971, researchers deliberately infected institutionalized intellectually disabled children with hepatitis at Willowbrook State School, an institution on Staten Island, New York, the experiments were described as "the most unethical medical experiments ever performed on children in the United States."[110] That condemnation came despite the fact that 90 percent of the children at Willowbrook were getting hepatitis anyway, and the research did lead to a better understanding of how hepatitis spreads. When such harmful experiments on human beings become known, they lead to an outcry against the experimenters, and rightly so. They are, very often, a further example of the arrogance of researchers who justify their work on the grounds of contributing to knowledge. Nevertheless, the contrast between this condemnation of experiments on members of our own species, and the acceptance of experiments that cause more extreme suffering for less benefit—but are performed on animals, not humans—is so marked that it defies rational defense. Even at the height of the coronavirus pandemic in 2020, there was considerable—in my view, excessive—hesitancy about accepting the offers of fully informed volunteers who were prepared to be infected with the virus in order to speed up the search for effective vaccines.[111] There was no hesitancy at all about using animals who cannot be informed and do not volunteer.

This blatant speciesism in research should evoke echoes of the way in which blatant racism has led to painful experiments on other races, defended on the grounds of usefulness for the dominant race. Under the Nazi regime in Germany, nearly two hundred doctors, some of them eminent in the world of medicine, took part in experiments on Jews and Russian and Polish prisoners. Thousands of other physicians knew of these experiments, some of which were the subject of lectures at medical academies. Yet the records show that the doctors sat through verbal reports describing horrendous injuries inflicted on people of these "lesser races," and then proceeded to discuss the medical lessons to be learned from them, without anyone making even a mild protest at the nature of the experiments. Then, as now, horrendous injuries were inflicted on research subjects, the results were written up in dispassionate scientific jargon and then were calmly discussed by researchers.[112] As the great Jewish writer Isaac Bashevis Singer has written: "In their behavior toward creatures, all men were Nazis."[113]

Experimentation on subjects outside the experimenters' own group is a story that constantly repeats itself with different victims. In the United States, the most notorious twentieth-century instance of human experimentation was the deliberate nontreatment of Black syphilis patients at the Tuskegee Institute, Alabama, so that the natural course of the disease could be observed. This was continued long after penicillin was shown to be an effective treatment for syphilis.[114] New Zealand's major human experimentation scandal came to light in 1987, when it was discovered that Dr. Herbert Green, a respected doctor at a leading Auckland hospital, had decided not to treat patients with early signs of cervical cancer. He was trying to prove his unorthodox theory that these early signs would not become invasive, but he did not tell the patients that they were part of an experiment. His theory was wrong, and twenty-seven of his patients died. The victims were, of course, all women.[115]

To return to the question of when an experiment might be justifiable: In extreme circumstances, absolutist answers always break down. Torturing a human being is almost always wrong, but it is not absolutely wrong. If torturing a terrorist were the only way in which we could discover the location of a nuclear bomb hidden in a New York City basement and timed to go off before the city could be evacuated, then torture would be justifiable. Similarly, if a single experiment could cure cancer, that experiment would be justifiable. But in real life the benefits are always more remote and uncertain. So how do we decide when an experiment is justifiable?

We have seen that experimenters reveal a bias in favor of their own species whenever they carry out experiments on nonhumans, where experimenting on human beings would never be justified, even on severely brain-damaged and orphaned ones. This principle suggests an answer to our question. Since a speciesist bias, like a racist bias, is unjustifiable, an experiment cannot be justifiable unless the experiment is so important that the use of a profoundly brain-damaged human would also be justifiable. We can call this the non-speciesist ethical guideline.

Some people think that all humans, irrespective of their mental capacities, have rights that must never be violated, no matter what the benefit. They think that to experiment on human beings in ways that harms them, without their informed consent, is always wrong. For them, the non-speciesist ethical guideline would also mean that to experiment on animals in ways that harm them is always wrong.

I accept the non-speciesist ethical guideline, but I do not think that it is *always* wrong to experiment on profoundly brain-damaged humans or on animals in ways that harm them. If it really were possible to prevent harm to many by an experiment that involves inflicting a similar harm on just one, and there was no other way the harm to many could be prevented, it would be right to conduct the experiment.

In 2006, during the filming of a BBC documentary on animal testing, I was confronted with what may be such a defensible example of research. Tipu Aziz, an Oxford University neurosurgeon specializing in parkinsonism—the condition characteristically associated with Parkinson's disease—sought my views on the ethics of his experiments on about a hundred monkeys that had led to significant improvements in the condition of 40,000 people with parkinsonism. I answered hypothetically that if there was no other way of discovering the knowledge used to make so many people better, this research would be justifiable. Some people thought that this was a softening of a hard-line position against animal research that I took in earlier editions of this book, but the sentence that you read a few pages back—"It will not do to say 'Never!'"—is already in the first edition.

In 2018, Aziz and two co-authors stated that 100,000 patients had by then been treated with various forms of deep brain stimulation developed as a result of research on monkeys and that the treatments have proved highly effective at treating a range of symptoms of Parkinson's disease, although they do not cure the condition. On the other hand, the number of monkeys used has presumably also risen, since the authors say that future developments will continue to require the use of primates due to their similarity to humans. The authors explicitly acknowledge something many researchers overlook: that this similarity "brings with it substantial ethical responsibilities on the part of the researchers."[116] I do not have the expertise to assess the claim that these developments could not have happened without using monkeys as experimental subjects, but if that is the case, and if suffering was minimized, then this research does appear to bring greater benefits than it causes harm, even when we give this harm no less weight than we would give to similar harm inflicted on a human being with cognitive capacities no greater than those of the monkeys used.

This takes us back to the problem of where to draw the line. As this chapter has shown, appalling suffering is being inflicted on millions of animals for purposes that in any impartial view are obviously inadequate to justify the suffering. These experiments obviously fall on the wrong side of the line. When we have ceased to carry out such experiments, then there will be time to discuss, in particular cases, on which side of the line an experiment falls.

In the United States, where the present lack of control over experimentation still allows for the kinds of experiments described in the preceding pages, a minimal first step would be a requirement that no experiment be conducted without prior approval from an ethics committee authorized to refuse approval to experiments when it does not consider that the potential benefits outweigh the harm to the animals. As we have seen, systems of this kind already exist in Australia, Sweden, and other countries, where they are accepted as fair and reasonable, not only by the general public but also by the scientific community. On the basis of the ethical arguments in this book, such a system falls far short of the ideal. The animal welfare representatives on such committees often come from groups that hold a spectrum of views, but, for obvious reasons, those who receive and accept invitations to join animal experimentation ethics committees tend to come from the less radical groups within the movement. They may not themselves regard the interests of nonhuman animals as entitled to equal consideration with the interests of humans; or if they do hold such a position, they may find it impossible to put it into practice when judging applications to perform animal experiments because they would be unable to persuade other members of the committee to agree with them. Instead, they are likely to insist on proper consideration of alternatives, genuine efforts to minimize pain, and a clear demonstration of significant potential benefits, sufficiently important to outweigh any pain or suffering that cannot be eliminated from the experiment. Most animal experimentation ethics committees operating today apply these standards in a speciesist manner, weighing animal suffering more lightly than potential comparable human benefit. Even so, an emphasis on such standards would eliminate many painful experiments now permitted and would reduce the suffering caused by others.

In a society that is fundamentally speciesist, there is no quick solution to such difficulties with ethics committees. For this reason some animal rights advocates will have nothing to do with them. Instead they demand the total and immediate elimination of all animal experimentation. Such demands have been put forward many times during the last century and a half of anti-vivisection activity, but they have shown no sign of winning over the majority of voters in any country. Meanwhile the greatest progress in reducing the suffering of animals in laboratories has come from the activist campaigns described earlier in this chapter, like those run by Henry Spira and People for the Ethical Treatment of Animals. These breakthroughs have resulted from the work of people who found a way around the "all or nothing" mentality that had effectively meant nothing for animals.

A WAY FORWARD?

SUPPOSE THAT WE WERE ABLE TO GO BEYOND MINIMAL REFORMS of the sort that already exist in the more enlightened nations. Suppose we could reach a point at which the interests of animals really were given equal consideration with the similar interests of human beings. That would mean the end of the vast industry of animal experimentation as we know it today. Around the world, cages would empty and laboratories would close down. It should not be thought, though, that medical research would grind to a halt or that a flood of untested products would appear on the market. So far as new products are concerned, we would make do with fewer of them, using ingredients already known to be safe, and there would be little hardship, and some benefits, in that. But for testing really essential products, as well as for other kinds of research, alternative methods not requiring animals, or at least not causing them significant suffering, can and would be found.

The defenders of animal experimentation are fond of telling us that we owe our increased life expectancy to research on animals.

In Britain, for example, in the midst of a debate over reform of the law on animal experimentation, the Association of the British Pharmaceutical Industry ran a full-page advertisement in the *Guardian* under the headline "They say life begins at forty. Not so long ago, that's about when it ended." The advertisement went on to say that it is now considered to be a tragedy if a man dies in his forties, whereas in the nineteenth century, when the average life expectancy was only forty-two it was commonplace to attend the funeral of a man in his forties. The advertisement stated that "it is thanks largely to the breakthroughs that have been made through research which requires animals that most of us are able to live into our seventies." This advertisement was so blatantly misleading that David St. George, a specialist in community medicine, wrote to the *Lancet* to say that "the advertisement is good teaching material, since it illustrates two major errors in the interpretation of statistics." The most obvious error is that a life expectancy of forty-two does not mean that death at that age is commonplace. When the life expectancy was forty-two, the most common age at death was during the first year of life, and anyone who survived childhood had a good chance of living into their seventies.

St. George drew on Thomas McKeown's influential book *The Role of Medicine*,[117] which set off a debate about the relative contributions of social and environmental changes, as compared with medical intervention, in improvements in mortality since the mid-nineteenth century. St. George wrote: "This debate has been resolved, and it is now widely accepted that medical interventions had only a marginal effect on population mortality and mainly at a very late stage, after death rates had already fallen strikingly."[118]

J. B. and S. M. McKinley reached a similar conclusion in a study of the decline of ten major infectious diseases in the United States. They showed that in every case except polio, the death rate had already fallen dramatically (presumably because of improved sanitation and diet) before any new form of medical treatment was introduced. Concentrating on the 40 percent fall in crude mortality in the United States between 1910 and 1984, they estimated "conservatively" that perhaps 3.5 percent of the fall in the overall death rate can be explained through medical interventions for the major infectious diseases. Indeed, given that it is precisely for these diseases that medicine claims most success in lowering mortality, 3.5 percent probably represents a reasonable upper-limit estimate of the total contribution of medical measures to the decline in infectious disease mortality in the United States.[119] Moreover this 3.5 percent is a figure for all medical intervention. The contribution of animal experimentation itself can be, at most, only a fraction of this minor contribution to the decline in mortality.

No doubt there are some fields of scientific research that will be hampered by consideration of the interests of animals used in experimentation. Important discoveries often mentioned by those defending animal experimentation go back as far as Harvey's work on the circulation of blood in the seventeenth century and include Banting and Best's discovery of insulin and its role in diabetes; the recognition that polio is caused by a virus and the development of a vaccine for it; several discoveries that served to make open-heart surgery and coronary artery bypass graft surgery possible; and the understanding of our immune system and ways to overcome rejection of transplanted organs.[120] The claim that animal experimentation was essential in making these discoveries has been denied by some opponents of experimentation.[121] I do not intend to go into this controversy here. We have just seen that any knowledge gained from animal experimentation has made at best a very small contribution to our increased lifespan; its contribution to improving the quality of life is more difficult to estimate.

In a more fundamental sense, the controversy over the benefits derived from animal experimentation is essentially irresolvable, because even if valuable discoveries were made using animals, we cannot say how successful medical research would have been if the immense resources currently being put into research on animals had gone into clinical research on humans and into developing alternative testing methods. Some discoveries would probably have been delayed, or perhaps not made at all; but many false leads would also not have been pursued, and it is possible that medicine would have developed in a very different and more efficacious direction.

In any case, the justifiability of animal experimentation cannot be settled by pointing to its benefits for us, no matter how persuasive the evidence in favor of such benefits may be. There is nothing sacred about the right to pursue knowledge. As we have seen, we already accept many restrictions on the scientific enterprise. We do not believe that scientists have a general right to perform painful or lethal experiments on human beings without their consent. The restriction on harmful experimentation without consent needs to be broadened beyond our own species, while at the same time the legitimacy of occasional, carefully scrutinized exceptions should be recognized, and there should be greater acceptance of well-informed, freely consenting volunteers as research subjects.

Finally, it is important to realize that the health problems that take the greatest number of human lives continue to exist, not because we do not know how to prevent disease and keep people healthy, but because few people are putting money into doing what we already know how to do (the Gates Foundation is the most notable exception). The diseases that ravage poorer parts of the world, like diarrhea, malaria, and pneumonia, are conditions that, by and large, we know how to prevent or cure. They have been eliminated in communities that have adequate nutrition, sanitation, and health care. It has been estimated that 5 million children under five die each year—that's 13,800 every day—and most of these deaths could be prevented if we worked harder to provide children in low-income countries with safe drinking water, bed nets that protect against malaria-carrying mosquitoes, better nutrition, and vaccinations.[122]

When all this has been said, there still remains the practical question: What can be done to change the widespread practice of experimenting on animals? Undoubtedly, some action that will change government policies is needed, but what action precisely? What can the ordinary citizen do to help bring about change?

Legislators tend to ignore protests about animal experimentation from their constituents because they are overly influenced by scientific, medical, and veterinary groups. In the United States, these groups maintain registered political lobbyists in Washington, and they lobby hard against proposals to restrict experimentation. Since legislators do not have the time to acquire expertise in these fields, they rely on what the "experts" tell them. But this is a moral question, not a scientific one, and the "experts" usually have an interest in the continuation of experimentation or else are so imbued with the ethic of furthering knowledge that they cannot detach themselves from this stance.

Nor is the task of reform made any easier by the large companies involved in the profitable businesses of breeding animals and selling them to research institutions, or manufacturing and marketing the cages for them to live in, the food used to feed them, and the equipment used to experiment on them. These companies are prepared to spend huge amounts of money to oppose legislation that will deprive them of their profitable markets. With financial interests like these allied to the prestige of medicine and science, the struggle to end speciesism in the laboratory is bound to be difficult and protracted. So what is the best way to make progress?

It does not seem likely that any major Western democracy is going to abolish animal experimentation at a stroke, but in 2021, the European Parliament passed a resolution calling on the European Commission to work with scientists to develop a plan to phase out experiments on animals. Achievement of that goal will, no doubt, lie some years in the future, but to have it endorsed by a parliament representing the people of twenty-seven countries that together conduct experiments on about 10 million animals every year is a sign of how far the debate in Europe has shifted in favor of taking the interests of animals seriously.[123]

It has taken many years of hard work to make animal experimentation a political issue in Europe. Fortunately this has now started to happen in several countries, where politically astute animal advocates—some of them with representatives in national parliaments—have sufficiently large followings to pressure the major political parties, whether right, center, or left, to develop policies on improving animal welfare. Then, when campaigning for election, these parties feel the need to make specific commitments that will attract voters.

The exploitation of laboratory animals is part of the larger problem of speciesism and it is unlikely to be eliminated altogether until speciesism itself is eliminated. Surely one day, though, our children's children, reading about what was done in laboratories in our era, will feel the same sense of horror and incredulity at what otherwise civilized people could do that we now feel when we read about the atrocities of the Roman gladiatorial arenas or the eighteenth-century slave trade.

CHAPTER 3

Down on the Factory Farm . . .

or what happened to your dinner when it was still an animal

HOW AMERICAN VETERINARIANS DELIBERATELY HEATED 243,016 PIGS TO DEATH

IN AUGUST 2021, FIVE VETERINARIANS PUBLISHED AN ARTICLE IN the *Journal of the American Veterinary Medical Association* that describes, in calm and precise language, and with charts and tables, how, between April and June 2020, they planned and directed the killing of 243,016 pigs from several states in the American Midwest. They arranged for the pigs to be transported to barns that had been specially retrofitted and sealed for this purpose. Workers locked the pigs, in groups of about 1,500 adults or 3,700 infant pigs, into each barn. The teams running the operation then shut off the ventilation and used heaters and a steam generator to raise the temperature inside the barns above 130°F (54°C), with at least 90 percent humidity. In some cases, temperatures reached as high as 170°F (76.7°C). The veterinarians recorded the length of time between the moment the pigs were locked in the barns and the moment the desired temperature was reached and, from there, the "time to silent"—that is, the time when the pigs stopped squealing or moving. They reported that for the majority of the pigs it took about 30 minutes for the temperature inside the barns to reach 130°F, and a further 65 minutes from that time until the pigs were silent and still, although one cycle took more than 2.5 hours. The infant pigs died slightly faster. This procedure was carried out 138 times, until all the pigs were dead.[1]

The pigs whose deaths the veterinarians directed all belonged to Iowa Select Farms, Iowa's largest producer of pigs, but other major U.S. producers have killed pigs in a similar manner. The veterinarians estimated that in the United States during 2020, one million pigs were, to use their euphemism, "depopulated." Why were these animals heated to death en masse? you may ask. At the time, the slaughterhouses to which the pigs would normally have been sent were closed because of the COVID-19 pandemic. The veterinarians say that there were no alternative locations for slaughter, nor for the pigs (whom they describe as being "housed in overstocked barns") to be kept in until the slaughterhouses re-opened. They say that other methods of euthanasia that inflict a quicker death than inducing heatstroke, such as the use of carbon dioxide, were simply not available.

Everyone who has lived through the COVID-19 pandemic will be familiar with the supply chain problems that occurred at that time. But the industry we are concerned with here produces sentient animals, not toilet paper. The producers have total control over the lives and deaths of these animals, and so they must take full responsibility for their fate. If a cruise ship sank in an unusually severe storm and the passengers all drowned because the ship's owner had not provided any lifeboats, we would not allow the owner to escape responsibility on the grounds that so severe a storm was unexpected. Similarly, it is wrong for producers of hundreds of thousands of animals not to plan for the possibility of something going wrong—and, when it does go wrong, to not have access to the trained staff, equipment, and materials required to implement that plan, whether it is alternative housing for the pigs or a humane way of killing them. What happened to these pigs throws a spotlight on the attitudes toward animals that underlie modern animal production. As we shall see in this chapter, corporations operating factory farms heated millions of animals to death both before the COVID-19 pandemic and afterwards as well, when slaughterhouses were fully operational. An industry that produces animals without planning for emergencies has demonstrated that it doesn't really care about the welfare of the sentient beings it is producing.

INTENSIVE ANIMAL PRODUCTION

FOR MOST HUMAN BEINGS, ESPECIALLY THOSE IN MODERN URBAN and suburban communities, the most direct form of contact with nonhuman animals is at mealtime: We eat them. This simple fact is the key to our attitudes about other animals, as well as the key to what each one of us can do about changing these attitudes. The use and abuse of animals raised for food far exceeds, in sheer numbers of animals affected, any other kind of mistreatment. Each year, in the United States alone, 34 million cows, 128 million pigs, 2 million sheep, 23 million ducks, 216 million turkeys, and 9.3 billion chickens are killed for food, for a total of 9.7 billion vertebrate land animals.[2] Global figures, of course, are far higher. The Food and Agriculture Organization of the United Nations estimates that more than 83 billion birds and mammals are slaughtered for food each year.[3] If it takes you two minutes to read a page of this book, approximately 316,000 of these individual sentient beings have been slaughtered during that time.

It is here, on our dinner tables and in our neighborhood supermarkets, that we are brought into direct touch with the most extensive exploitation of other species that has ever existed. In general, we are ignorant of the abuse of living creatures that lies behind the food we eat. Buying food in a store or restaurant is the culmination of a long process, of which all but the end product is delicately screened from our eyes. We buy our meat and poultry in neat plastic packages. It hardly bleeds. There is no reason to associate this package with a living, breathing, walking, suffering animal. The very words we use conceal its origins, at least for the larger mammals: We eat "beef," not "bull," "steer," or "cow"; and "pork," not "pig"—although for some reason we seem to find it easier to face the true nature of a leg of lamb, and we don't need to disguise our consumption of chickens and fish. The term "meat" is itself deceptive. It originally meant any solid food, not necessarily the flesh of animals, and only in the fourteenth century was it commonly used for what had hitherto been called "flesh-meat." For another century, vegetables could still be called "green-meat" and some dairy foods "white-meat." Interestingly, the French *viande* originally meant "food," and also, over time, became limited to the flesh of animals when used as food.[4] By using the more general "meat" we avoid facing the fact that what we are eating is really "flesh." In several countries, the meat industry is now seeking to prevent anything that does not come from a dead animal being labeled "meat." It would be more accurate if they advertised their product as "flesh-meat" and other forms were labeled "plant-meat" or "cultured meat."

These verbal disguises are merely the top layer of a much deeper ignorance of the origin of our food. Consider the images conjured up by the word "farm": a house; a barn; a flock of hens, overseen by a strutting rooster, scratching around the farmyard; a herd of cows being brought in from the fields for milking; and perhaps a sow rooting around in the orchard with a litter of squealing piglets running excitedly behind her.

Few farms were ever as idyllic as that traditional image would have us believe, but traditional farms were at least aesthetically pleasing, and far removed from an industrial, profit-driven city lifestyle. When the first edition of this book was published in 1975, not many people knew much about modern methods of animal raising. The general belief—which I shared, until I was twenty-four years old—was that farm animals enjoy the natural pleasures of animal existence without the hardships that wild animals must endure in their struggle for survival.

Today, more people are aware that these comfortable assumptions bear little relation to the realities of modern farming. For a start, farming is no longer predominantly the work of independent farming families. During the last seventy years, large corporations and assembly-line methods of production have turned agriculture into agribusiness. Today, in the United States, the four biggest companies control 54 percent of the poultry industry; for pork, the top four companies hold 70 percent of the market; and in beef, the four biggest companies have 85 percent of the market.[5] In egg production, where a century ago a big producer might have had three thousand laying hens, today close to 90 percent of all U.S. eggs come from just 67 producers, each with more than half a million hens, and the largest have between 8 and 47 million.[6] The big corporations and those who must compete with them are not concerned with a sense of harmony among plants, animals, and nature. Producing animal products is competitive, and the methods adopted are those that cut costs and increase production. The giant companies dictate terms to the remaining producers, leaving them with few choices if they don't want to go bankrupt.[7] Animal production has become an industrial process, popularly known as factory farming, in which animals are treated like machines that convert low-priced fodder into more profitable flesh, and any innovation will be used if it results in a cheaper conversion ratio. Most of this chapter is a description of these methods, and of what they mean for the animals to whom they are applied.

Once again, however, my point is not that the people who do these things are especially cruel and wicked. On the contrary, the farming methods I am about to describe are merely the logical application of the dominant attitudes and prejudices that are discussed elsewhere in this book. Once we place nonhuman animals outside our sphere of moral consideration and treat them as things we use to satisfy our own desires, the outcome is predictable. It was well described by Ruth Harrison, the author of *Animal Machines*, a pioneering exposé of intensive farming methods in Britain: "Cruelty is acknowledged only where profitability ceases."[8] In other words, as I shall prove with several examples in the pages to come, the welfare of the animal counts only when, if it got any worse, that would reduce profits.

Running a profitable animal production system is compatible with—and in a competitive marketplace may even require—appalling amounts of animal suffering. Especially in the United States, industrial animal producers have resisted change as strongly as they can, spending large sums in attempts to defeat legislation intended to improve the conditions of farmed animals. And when they lose at the ballot box, they go to court to try to block the legislation.[9]

In order to make my account as objective as possible, I have not based the descriptions that follow on my own personal observations of farms and the conditions within them. Instead, what follows is drawn largely, though not entirely, from the sources that can be expected to be the most favorable to the farming industry: the websites, magazines, and trade journals of the farm industry itself, supplemented with peer-reviewed studies from scientists specializing in the relevant fields of animal production.

THE AGRIBUSINESS CHICKEN

THE FIRST ANIMAL TO BE REMOVED FROM THE RELATIVELY NATURAL conditions of the traditional farm was the chicken. Human beings use chickens in two ways: for their flesh and for their eggs. There are now standard mass-production techniques for obtaining both of these products. Promoters of agribusiness consider the rise of the chicken industry to be one of the great success stories of farming. At the end of World War II, chicken for the table was still relatively rare. It came mainly from small independent farmers or from the unwanted males produced by egg-laying flocks. Today in the United States, more than 8 billion "broilers"—as chickens bred to be eaten when young are called—are slaughtered every year.[10] Each of these birds is an individual social animal, with a distinct personality. As Konrad Lorenz, a pioneer in the science of animal behavior, wrote in the days when flocks were still small and farmers knew each of their birds:

> Do animals thus know each other among themselves? They certainly do. . . . Every poultry farmer knows that . . . there exists a very definite order, in which each bird is afraid of those that are above her in rank. After some few disputes, which need not necessarily come to blows, each bird knows which of the others she has to fear and which must show respect to her. Not only physical strength, but also personal courage, energy, and even the

> self-assurance of every individual bird are decisive in the maintenance of the pecking order.[11]

The essential step in turning chickens from farmyard birds into manufactured items was confining them indoors. Typically, a producer contracts with one of the big poultry corporations, which delivers 50,000 or more day-old chicks from the hatcheries, along with their feed and instructions on how to raise them. The producer puts the birds into huge, windowless sheds, holding 20,000 to 30,000 birds each. These chicks have been selectively bred to have voracious appetites and to grow rapidly. Inside the shed, every aspect of the birds' environment is controlled to make them grow faster on less feed. Food and water are supplied automatically. The lighting is adjusted according to advice from agricultural researchers: For instance, there may be bright light for twenty-three or even twenty-four hours a day for the first week or two, to encourage the chicks to gain weight quickly; then there may be a longer period of darkness, perhaps up to six hours a day. As the chickens grow and become more crowded, the lights will be kept dim at all times to reduce the aggression caused by crowding. For the last few days before the birds are rounded up and taken away to be killed, the period of darkness will again be reduced, to encourage more eating and hence more weight gain.[12]

Broiler chickens are killed when they are between five and seven weeks old—usually closer to five weeks in Europe, where a smaller bird is preferred, and closer to seven in the United States.[13] (The natural life span of a chicken is about seven years.) At the end of this brief period, the birds weigh, on average, 6.5 lbs, or just under 3 kg. This is an astonishing rate of growth for a chicken less than seven weeks old: In 1925, it took the average chicken 16 weeks to reach a weight of 2.5 lbs.[14] As the birds grow, they take up more space, and the shed becomes more crowded, until the birds have as little as half a square foot of space per chicken—or less than the area of a sheet of standard typing paper.

In the United States, the National Chicken Council recommends 0.7 square feet per bird, but the council's recommendations are just that—recommendations, with no legal authority. And even in countries that do have legally binding maximum stocking densities, producers may exceed them. A Canadian study of the effects of stocking densities on productivity notes, in passing, that although the maximum stocking density in Canada is 38 kilograms per square meter, the commercial producer whose stocking densities they had been invited to study—and therefore not a rogue producer operating in secret but one that welcomed outside researchers—sometimes had densities of more than 41 kilograms per square meter, or just over half a square foot per bird.[15] In the European Union stocking densities are also legally regulated, but can permit as many as twenty-five birds, weighing 2 kg each, on a square meter, which is 0.43 square feet per bird.[16] Higher stocking density means bigger profits. As an industry manual put it: "Limiting the floor space gives poorer results on a per bird basis, yet the question has always been and continues to be: What is the least amount of floor space necessary per bird to produce the greatest return on investment?"[17]

If you walk into a chicken shed, the first thing you are likely to notice is an overpowering stink; and soon you realize that it isn't just a harmless stink either; the air is stinging your eyes, which may start to tear up, and burning your lungs.[18] It's ammonia, from the birds' droppings, which are typically allowed to accumulate in the litter on the floor, not just for the six or seven weeks that one flock is there but often—to save the expense of new litter and the burden of disposing of the old litter—for years. According to extension poultry scientists at the University of Georgia, "many poultry companies and broiler producers have adopted the practice of reusing litter for one, two or even more years of production." In fact, they add, this has "become a standard in the industry." The U.S. Environmental Protection Agency sets maximum levels for human exposure to ammonia in the air: 35 parts per million (ppm) for 15 minutes exposure, and 25 ppm for exposure up to 8 hours. But the University of Georgia scientists refer to levels of 50–110 ppm, and the chickens are confined in the shed, living in that air, every day, for their entire lives. They develop chronic respiratory disease from breathing the ammonia-laden air, and because ammonia is corrosive, they often develop sores on their hocks and blisters on their breasts if they rest on litter that has absorbed moisture. At the highest levels of exposure, birds may go blind.[19]

Once, birds who fell sick or were injured could be attended to, or if necessary, quickly killed. Now one person looks after tens or even hundreds of thousands of birds—except that with so many birds, there really isn't any "looking after" at all. As long ago as 1970, the U.S. secretary of agriculture wrote enthusiastically about how the then-new method of raising chickens indoors enabled one person to "care" for 60,000 to 75,000 chickens.[20] The same idea of "caring" for an impossibly large number of birds is still in use. Under the heading "animal welfare," a 2018 article in *Farm and Dairy*, an Ohio-based farming newspaper, describes Hannah Malmsberry as "nurturing" more than 100,000 chickens. She clearly isn't looking at the birds individually, for to spend even one second a day inspecting each bird would take more than twenty-four hours.[21] University of Georgia poultry scientists have said that the labor requirements of modern chicken houses amount to between four and six minutes per 1,000 birds, which works out to about one-third of a second for each bird.[22]

One of Hannah Malmsberry's main tasks, the *Farm and Dairy* article acknowledged, is removing the corpses of the 3 to 4 percent of birds who don't even survive the forty-five days it takes them to grow to market weight. That's 3,000–4,000 dead chickens, or about seventy-seven every day, but it's still better than the industry as a whole, which in 2021 had an average mortality of 5.3 percent.[23] In that year, according to official figures, the number of chickens that did make it to slaughter in the United States was at least 9,210,889,000.[24] That means that the producers started out with at least 9,726,387,539 chickens, and more than half a billion of them died in the first seven weeks of their lives.

Chickens in broiler houses succumb to a number of diseases related to their rapid growth. One common cause of death is known by a variety of names: "Sudden Death Syndrome," "Flip-over Disease," and "Dead in Good Condition." As described in a veterinary manual, affected birds "die suddenly, with a short, terminal, wing-beating convulsion." Many affected broilers just "flip over and die on their backs."[25] How common this is depends on the conditions in the broiler house and on the strain of birds, but one study found its incidence in broiler chickens to range from 0.5 to 9.62 percent. Because producers want fast growth, they will accept the deaths of a percentage of their birds rather than slow down the speed at which their birds put on weight.[26]

If human babies grew at the same rate as today's chickens, researchers at the University of Arkansas have pointed out, by the time they were two months old, they would weigh 660 pounds, or 300 kilograms![27] This rate of growth leads to other problems that are, from the point of view of animal welfare, of more concern than sudden death. After all, the birds who die prematurely in the broiler house are really the lucky ones—they are spared the trauma of being gathered up, crammed into trucks, and driven to the slaughterhouse. Less fortunate are the birds in whom the fast growth rate causes direct and prolonged suffering, most obviously because the birds' immature leg bones struggle to support the weight of their bodies. Already in the early 1990s, a study found that 90 percent of broilers had detectable leg problems, while 26 percent suffered chronic pain as a result of bone disease. The researchers linked these problems to genetic selection for fast growth, but that selection has continued over the past thirty years, so that nowadays growth is even faster.[28] Professor John Webster of the University of Bristol's School of Veterinary Science, one of the most highly respected experts on the welfare of farmed animals, has said: "Broilers are the only livestock that are in chronic pain for the last 20 percent of their lives. They don't move around, not because they are overstocked, but because it hurts their joints so much."[29] The birds may try to avoid the pain by sitting down, but they have nothing to sit on except the ammonia-laden litter, which, as we saw earlier, is so corrosive that it can burn their bodies. Their situation has been likened to that of someone with arthritic leg joints who is forced to stand up all day. Webster has described modern intensive chicken production as "in both magnitude and severity, the single most severe, systematic example of man's inhumanity to another sentient animal."[30]

Webster was referring to the suffering that is bred into the system of intensive chicken production; but videos taken by undercover investigators show that sometimes the suffering is made worse still by the callousness of those who have total control over the lives and deaths of the chickens. In 2014, for example, an investigator working for the animal advocacy organization Compassion Over Killing went undercover with a North Carolina producer growing chickens for Pilgrim's, one of America's biggest brands. The video shows live but sick or disabled chickens being picked up and thrown into plastic buckets. When the buckets are full they are taken outside and emptied down a hatch. "We have an open pit," explains one of the workers. "It's a deep ravine, it's got pallets on it, like doorways. We dump them in there and

Mother Nature takes care of the rest of them." He adds, with a laugh: "You look down there and it's like a gravy, just simmering and squirming." The video shows a worker closing the hatch on top of chickens still able to walk and flap their wings. The investigator asks, "This one's still alive, should we hit it?" and receives this response: "No, we're going to drop them in the pit, just like they are." He goes on to say that it is possible to kill them by wringing their neck, then adds: "But I mean, if you got buckets and buckets, you're going to stand here all day and . . . [He demonstrates the neck-wringing motion with his hands, shrugs his shoulders, and continues.] You're supposed to do it, but . . ."

Pilgrim's will say, of course, that once they learned of this, they terminated their relationship with the grower. But we don't know for how many years those practices had been going on before the courageous undercover investigation exposed them, or how many other producers use equally cruel ways of disposing of their disabled chickens. Nor is the $31 billion a year chicken industry eager to find out. Instead, they lobby for "ag-gag" laws, which make it a crime to take photos or video on agricultural property without the permission of the owner.[31]

Each time the appalling suffering inflicted on chickens produced for a major brand is exposed, the company that owns the brand says it will take strong action to prevent such abuses. And yet we see it happen again and again. So, predictably, in February 2022, as I was revising this chapter, yet another set of videos was released, this time taken on a facility in Italy that supplies chickens to AIA, the fourth-largest producer in Europe. The most striking feature of the video is the many birds with deformities that make them unable to walk or to reach the drinking trough. They lie on the ground flailing helplessly, or struggling to move but succeeding only in turning themselves around in circles on the litter. A statement about the investigation released by the Spanish animal protection organization Equalia, working together with The Humane League UK, points out that many of the problems depicted in the video are referred to in the industry as "production diseases"—as if they are somehow justified because they are part of genetically selecting birds to have huge appetites and gain weight as fast as possible.[32]

Some of the birds in the video may have been suffering from spondylolisthesis, or as it is more widely known in the chicken industry, "kinky back," known to be influenced by "genetic background and growth rate." A leading veterinary manual recommends culling these birds "because they are not ambulatory, unable to access food and water, and will die of starvation or dehydration."[33] But if, as we have seen, one person is responsible for the care of 100,000 birds (and unlikely to be keen to spend more time in the ammonia-filled atmosphere inside the sheds), it is quite likely that these birds will die slowly rather than being put out of their misery.

Defenders of factory farming often say that if the birds or animals were not happy, they would not thrive and hence would not be profitable. The grim truth about the chicken industry provides a clear refutation of this naive myth. Everyone knows how to stop the suffering of the chickens, but it continues *because* it is profitable. Each death before the chickens reach the slaughterhouse costs the grower money, but the bottom line is determined by how much saleable product each batch of birds produces. Researchers at the University of Arkansas set out to discover if producers would make more money by raising a slower-growing breed of birds with lower mortality. They found that the answer depends on the precise costs involved but concluded that often "it is better to get the weight and ignore the mortality."[34]

The suffering caused by breeding for rapid growth is not limited to the young birds sold to be eaten. The parents of the chickens sold in supermarkets, known in the industry as "breeder birds," are, of course, bred to have the genes that are desired in their offspring—genes that give them their insatiable appetites and ability to convert food to body mass. As we have just seen, this causes widespread leg problems in the young chickens raised for meat, but, despite the chronic leg pain those birds experience, most of them survive until they are slaughtered at six or seven weeks old. If they were not killed, and given all the food they wanted, they would gain even more weight and many more would die with each passing week. Chickens are not able to breed, however, until they are eighteen to twenty weeks old, and once they reach that age the producers want them to keep breeding as long as possible. If they were able to eat as much as they wanted, they might well die before they became sexually mature; and if they did reach maturity, they would be so fat that they would not be able to mate. The chicken industry has a characteristic solution: to give the parents 60 to 80 percent less than they would eat if the food were available.[35]

In an article on feeding the genetic parents of fast-growing chickens, poultry specialists at the University of Georgia described how those who raise the breeding birds in the United States use a "Skip-a-Day (SAD) feeding program," which means that the birds are fed only every second day.[36] A summary by Canadian scientists of several studies on the effect of these diets shows that the abbreviation is appropriate. Birds on SAD feeding spend 25 percent of their time pecking at the empty feeder. They also start eating their litter, which consists of indigestible substances such as wood shavings and feathers. Their unsatisfied desire to eat is so strong that it can lead to aggression and what the researchers described as "redirected oral behavior toward themselves, other pullets and equipment or objects."[37] In other words, these birds are so desperately hungry that they will try to eat anything.

In 2017 the U.S. National Chicken Council, an industry body, issued a document, available on its website, titled "National Chicken Council Animal Welfare Guidelines and Audit Checklist for Broiler Breeders."[38] Under the heading "Nutrition and Feeding" it says: "Decisions to modify the rate of weight gain (for production or health related reasons) are acceptable. Moderation of feed intake may be used to maintain correct frame size, body composition, and weight gain." The guidelines do show an awareness that "moderated feeding programs may result in overconsumption of water, which can adversely impact welfare." But the "acceptable" remedy is not providing more food; instead it is to restrict the birds' water intake as well. If the guidelines really were based on concern for the welfare of the chickens, they would say: "Choose only those breeds of chickens in which the birds grow at a rate that allows their genetic parents to reach sexual maturity and reproduce without being hungry for most of their lives." But the National Chicken Council's animal welfare guidelines never say that.

I will leave the last word to the British judge in a case that became known as "the McLibel trial." In 1997 McDonald's sued two British environmental activists, Helen Steel and David Morris, who had distributed a leaflet accusing the burger giant of, among other things, being responsible for cruelty to animals. After examining the evidence presented by Steel and Morris, the judge concluded that the claim that McDonald's was responsible for cruelty was not defamatory because it was true. This verdict was achieved despite the fact that McDonald's was able to afford a top legal team, while Steel and Morris represented themselves. With regard to the plight of the breeding birds, the judge said: "My conclusion is that the practice of rearing breeders for appetite, that is to feel especially hungry, and then restricting their feed with the effect of keeping them hungry, is cruel. It is a well-planned device for profit at the expense of suffering of the birds."[39]

If living in long, crowded, ammonia-filled, dusty, windowless sheds is stressful, the birds' first and only experience of sunlight is even worse. When the animals are said to be ready for slaughter, workers come in and grab them by the legs, carry them out upside down, and summarily stuff them into crates, which are piled onto the back of a truck. To reduce the damage done to the birds, which has an economic cost, some agricultural equipment manufacturers have promoted mechanized systems as a better option. Peer System, a Dutch manufacturer, says: "Catching poultry is in many cases still done by hand, which makes it an intensive, heavy task for employees and creates very stressful situations for the animals. They panic and easily get damaged." That's all true. Nevertheless, viewing Peer System's own promotional video of their mechanized alternative in operation shows that this must still be a very stressful experience for the birds, who are swept up by a large machine onto a conveyor belt which then dumps them into a container that is slotted into the back of a truck to be taken to the slaughterhouse. Peer System boasts that its catching system can move 9,000 chickens per hour.[40]

At the slaughterhouse, euphemistically called a "processing plant," the birds are taken off the truck and usually hung upside down, with their feet in shackles, on a moving conveyor belt. Chickens have pain receptors in their legs, and to hang them, fully conscious, upside down by their legs in a metal shackle is already a painful experience. The conveyor belt then takes them to a blade that will cut their throat. In most industrialized countries they are supposed to be stunned before that happens. The most common method of stunning, both in the United States and Europe, is for the conveyor belt to dip the birds into an electrified water bath. If the electric current is set too high, however, the birds may experience severe muscle contractions that adversely affect the quality of the meat; but if the current is too low, the birds will be paralyzed but not stunned. With the line between too high and too low being quite fine, and the commercial incentives in favor of ensuring that the meat is not spoiled, it is likely that many birds are not properly stunned. Some will have their throats cut while fully conscious, but others will also miss the blade. The conveyor belt then takes them into a tank of scalding water, intended to remove their feathers. Those birds that have not yet been stunned nor had their throats cut will die by drowning in scalding water.[41]

The plucked and dressed bodies of the chickens will then be sold to millions of families at an incredibly cheap price and served on the dinner table night after night. Rarely do consumers stop to ask themselves how a once-luxury food could become so cheap. And if they did stop to ask, where would they find the answer? The U.S. National Chicken Council declares: "Consumers want to be sure that all animals being raised for food are treated with respect and are properly cared for during their lives." The council then boasts about an audit that its welfare guidelines have received from the "Professional Animal Auditor Certification Organization, Inc." This means, it says, that "consumers and customers can feel confident that when buying and eating chicken, the birds were well cared for and treated humanely."[42] You may judge what that audit is worth from the fact that, as we have just seen, the council's guidelines allow its members to use breeds of chickens that can exist only if the parent birds are half starved in the first place. Apart from that, it's significant that even in its own media release, the council does not claim that the auditor examined the actual conditions of the chickens. It merely audited the council's guidelines. There is, therefore, no way in which it can provide assurance to consumers that the birds they are eating were actually humanely treated. If the council were truly interested in providing assurance about the welfare of commercially reared chickens, it would ask one of the respected and long-established national animal welfare organizations to make unannounced visits to chicken producers and then decide if the birds are being treated humanely.

In 2021 KFC, the international restaurant chain that operates in 150 countries, sought to improve the image of the chicken it serves by making a paid promotional film titled "Behind the Bucket" with YouTube influencer Niko Omilana. The film, which has been viewed more than a million times, shows Niko visiting the operations of Moy Park, a leading European chicken producer that sells chickens to KFC. At the time the film was made, the birds were very young, they looked cute, and there was still a reasonable amount of room in the shed. In front of the camera, Tony, the unit's manager, takes a bale of straw and says that they sprinkle fresh straw on the ground for the birds to play with and sit on. There are also various kinds of enrichment, including hanging perches. Niko asks Tony: "So you really do care about these chickens, then?" and gets the answer: "Yeah, we do." Niko expresses surprise that there are no cages and asks whether the chickens are fed steroids to make them grow faster. But no one in the animal welfare movement claims that chickens raised for meat are kept in cages—that applies to the egg industry—nor that they are fed steroids. The rapid growth of modern chickens is, as we have seen, from selective breeding.

After the film was released investigators for the vegan food brand VFC Foods made an uninvited visit to the same production facility. A video they made shows that most of the perches had disappeared, hoisted up to the ceiling where the chickens couldn't get at them—the investigators say that there was one perch left for the 52,000 birds in the shed. The video shows a floor that does not have any fresh straw—an investigator describes it as a mixture of woodchips and excrement. The birds are older than those Niko saw, and several have open sores on their skin from sitting on the urine-soaked litter. The video shows dead birds inside the barns, as well as birds who will soon die because their immature skeletons are unable to bear the weight of their fast-growing bodies, and they can no longer stand up or walk to food or water.

After the VFC Foods video was released and gained media attention, a spokesperson for Moy Park said that staff visit the sheds at least three times a day but acknowledged that "a small number of birds may die between those checks and are identified and removed during the next inspection." A spokesperson for the agency that represents Niko Omilana stated: "The filming that our client took part in accurately reported the conditions he saw and experienced on the day." That is no doubt true, but it merely shows how easy it is for public relations exercises to mislead people who do not insist on an unannounced inspection and do not know enough about the industry to ask the right questions. Niko Omilana has since taken the film down from his own social media channels.[43]

The reader who, after learning about the chicken industry in these pages is contemplating buying turkey instead of chicken, should be warned that this traditional centerpiece of the family's Thanksgiving dinner is now reared by the same methods as broiler chickens. Newly hatched turkeys are raised in incubators and then, before they are sent to the producers to be raised, they are debeaked and undergo partial toe amputation; the snood—the fleshy erectile protuberance that grows from the forehead of a male turkey—is also cut off. The cutting of the end of the beak and the talons is, as with chickens, to prevent the cannibalism that might otherwise occur in the crowded sheds in which they will live the rest of their lives, and the snood is removed because it is often a target for pecking from other birds. All this is done without any kind of pain relief. Turkeys stay in the sheds for four to five months, about three times as long as chickens, and their droppings accumulate on the floor for the entire time, so that the air reeks of ammonia.

Turkeys have even more leg problems than chickens because they are also bred to grow very rapidly, but in addition, the most popular breed of American turkey today, the aptly named "broad-breasted white," has been bred, as you can imagine, to have large breasts. These birds have been described as "physiologically unbalanced." They walk or stand less than older breeds, presumably because it is painful for them to put weight on their legs. A study of turkeys at thirteen different slaughterhouses found that 60 percent had swelling of the footpad and 41 percent had severe footpad dermatitis, while 25 percent had arthritis. These birds are likely to have been in pain. In addition, 30 percent of the turkeys had "breast buttons"—that is, blisters or other sores on the skin around their breastbone, which are common in turkeys who spend most of their time lying down on their sternum. So, for the misbegotten turkeys, there is no escape from pain. If they lie down to avoid the pain of carrying their heavy, unbalanced bodies on their arthritic legs and swollen feet, they end up with painful sores on their breastbone.[44]

Physical pain is not the only consequence of breeding turkeys to have huge breasts. The other outcome is that they are unable to have sex. The male's huge breast gets in the way. This raises a question that Americans might pose to the family around the Thanksgiving table: If they are eating a broad-breasted white, how was that turkey conceived? The answer: by artificial insemination.

In 2006, I co-authored a book titled *The Ethics of What We Eat*. My co-author, Jim Mason, had grown up on a farm in

Missouri. He noticed that the Butterball turkey company was advertising for people to work for its artificial insemination crew in Carthage, Missouri. Curious, he decided to apply, and as passing a drug test was the only requirement, he was accepted. He was told to catch the male turkeys by the legs and hold them upside down while another worker masturbated them. When the turkey ejaculated, Jim's co-worker used a vacuum pump to collect the semen in a syringe. This routine was repeated many times, until the semen, diluted with an "extender," filled the syringe. It was then taken over to a separate building, the "hen house," where the female breeding turkeys were kept.

For a time, Jim also worked in the hen house, which he described like this:

> You grab a hen by the legs, trying to cross both "ankles" in order to hold her feet and legs with one hand. The hens weigh 20 to 30 pounds and are terrified, beating their wings and struggling in panic. They go through this every week for more than a year, and they don't like it. Once you have grabbed her with one hand, you flop her down chest first with the tail end sticking up. You put your free hand over the vent and tail and pull the rump and tail feathers upward. At the same time, you pull the hand holding the feet downward, thus "breaking" the hen so that her rear is straight up and her vent open. The inseminator sticks his thumb right under the vent and pushes, which opens it further until the end of the oviduct is exposed. Into this, he inserts a straw of semen connected to the end of a tube from an air compressor and pulls a trigger, releasing a shot of compressed air that blows the semen solution from the straw and into the hen's oviduct. Then you let go of the hen and she flops away.

Workers in the hen house were told to do this to a hen in twelve seconds, so that they could get through 300 hens an hour—that's 3,000 in a standard 10-hour day. If Jim didn't keep up the pace, he got torrents of verbal abuse from the foreman while dodging shit spurting from the panicked hens. It was, he said, "the hardest, fastest, dirtiest, most disgusting, worst-paid work I have ever done." Under those conditions, for most of the abused and poorly paid workers, handling the birds gently would not be the first priority.[45]

CAGED HENS

"A HEN," SAMUEL BUTLER ONCE WROTE, "IS ONLY AN EGG'S WAY OF making another egg." Butler, no doubt, thought he was being funny; but when Fred C. Haley, president of a Georgia poultry firm, describes the hen as "an egg-producing machine" his words have more serious implications. To emphasize his businesslike attitude, Haley adds, "The object of producing eggs is to make money. When we forget this objective, we have forgotten what it is all about."[46]

Nor is this only an American attitude. A British farming magazine once told its readers:

> The modern layer is, after all, only a very efficient converting machine, changing the raw material—feedingstuffs—into the finished product—the egg—less, of course, maintenance requirements.[47]

The idea that the hen is just an efficient way to turn feed into eggs isn't good for hens, who are, unlike any machines we have invented so far, capable of feeling pleasure and pain. When people think of hens as machines, all the emphasis goes into making the hens do their job even more efficiently, no matter the cost to them.

The suffering of chickens bred for their eggs begins early in life. The fluffy little newly hatched chicks are immediately sorted into males and females by a "chick-puller." Since the male chicks don't lay eggs or grow fast enough to be reared for meat, they are an unwanted byproduct with no commercial value. Some companies gas them, but more often they are thrown, alive, into a machine that grinds them up between rotating drums. This is the fate of about 300 million chicks in the United States alone, and a similar number in the European Union, with the world total in the billions.[48] Thanks to the work of animal advocates, however, this mass killing of male chicks is now illegal in France (though with some exceptions) and Germany, and it will become illegal in Italy in 2026. Instead, machines that can detect the sex of the chick while in the egg are used, and the eggs containing males are discarded at an age at which the embryo is insufficiently developed to feel pain.[49]

Life for the female laying birds is longer than for the males, but whether this is a benefit is questionable. About eight days after hatching, she will have the sharp end of her beak cut off to prevent the dominant birds pecking at the weaker birds, who, in the crowded cages in which most U.S. hens are kept, are unable to move away. If the pecking draws blood, other birds may join in and may kill and eat the weaker birds. Providing the hens with more space and a richer environment, and selecting for less aggressive breeds, could eliminate the problem, but of course at a cost for the producers. The alternative solution is to cut back the hens' beaks, so that even if they peck, they will not draw blood. When this was first done, in the 1940s, the farmer would burn away the upper beaks of the chickens with a blowtorch. A modified soldering iron soon replaced that crude technique, and this was in turn replaced by a specially designed guillotine-like device with hot blades. After billions of chickens have, over many years, had the ends of their beaks cut off with hot blades, such techniques are now being replaced by an infrared device, which is more precise and does not leave open wounds.[50] But hens often then have their beaks cut again, between five and eleven weeks old, and this second cutting is usually done with a hot blade.[51]

"Debeaking" used to be the standard term in the industry for this procedure. (In China, where there is less resort to euphemisms regarding the treatment of farm animals, the literal translation of the term most commonly used is "cut mouth.") The most accurate term, used in some scientific papers, is "partial beak amputation."[52] Today the industry favors "beak trimming," which brings to mind painless procedures like trimming your hair or toenails. But if that is the impression that the industry is seeking to create, it is misleading. As an expert British government committee under zoologist Professor F. W. Rogers Brambell reported: "Between the horn and the bone is a thin layer of highly sensitive soft tissue, resembling the 'quick' of the human nail. The hot knife used in debeaking cuts through this complex of horn, bone, and sensitive tissue, causing severe pain."[53]

The damage done to the bird by debeaking is long term: Chickens mutilated in this way eat less and lose weight for several weeks.[54] The most likely explanation for this is that the injured beak continues to cause pain as the damaged nerves grow again, forming a mass of intertwining nerve fibers, called a neuroma. These neuromas have been shown in humans with amputated stumps to cause both acute and chronic pain. J. Breward and M. Gentle, researchers at the Poultry Research Centre of the British Agricultural and Food Research Council, found that this is probably also the case in the neuromas formed by debeaking.[55]

Another group of researchers examined many prior studies of

pain in chickens and found that debeaking leads to rapid changes in heart rate, breathing, and blood pressure, and it increases the stress hormones circulating in chickens. They confirm that, after debeaking, "chickens show prolonged changes in general activity, locomotion, feeding, drinking, preening, and pecking at objects; such changes can persist for many weeks depending on trimming age, method, and severity." The altered behavior that occurs following debeaking, they then say, is reduced by giving the birds a local anesthetic or such pain-relieving drugs as opioids or anti-inflammatories—but in commercial production, chickens receive no pain relief.[56]

Subsequently Gentle, expressing himself with the caution to be expected from a poultry scientist writing in a scientific journal, said:

> In conclusion, it is fair to say that we do not know how much discomfort or pain birds experience after beak trimming but in a caring society they should be given the benefit of the doubt. To prevent cannibalism and feather pecking of poultry, good husbandry is essential and in circumstances where light intensity cannot be controlled the only alternative is to attempt to breed birds which do not exhibit these damaging traits.[57]

When the first edition of this book appeared, all the big egg producers, worldwide, were keeping their laying hens in bare wire cages, too small for even a single hen to fully stretch or flap her wings—and there wasn't just one hen to a cage, but four or five. The reason for the crowding was the usual one: It pays. A carefully controlled study by members of the department of poultry science at Cornell University confirmed that crowding increases death rates. Over a period of less than a year, mortality among layers housed three to a twelve-by-eighteen-inch cage was 9.6 percent; when four birds were put in the same cage, mortality jumped to 16.4 percent; with five birds in the cage, 23 percent died. Despite these findings, the researchers advised that under most conditions the hens should be housed at four birds per cage, since the greater total number of eggs obtained made for a larger return on capital and labor, which more than compensated for the higher costs of what the researchers termed "bird depreciation."[58] Indeed, if egg prices are high, the report concluded, "five layers per cage make a greater profit."

In the cage system that has dominated the industry since the 1960s and is still the way most eggs are produced in the United States today,[59] hens can never walk around freely or lay their eggs in a nest. Yet until the 1970s, despite the fact that, in both the United States and Europe, hundreds of millions of hens were crowded into the cages, no major animal welfare organizations were campaigning against the cage system. That began to change in the late 1970s, with the rise of a new and more radical animal movement that used the media to show consumers where their eggs were coming from.

In Britain and several European countries, animal welfare was becoming a political force that politicians could no longer ignore. By documenting the conditions in factory farms, Compassion in World Farming, a tiny organization when it was founded in the early 1970s, began to gain broad public support, first in Britain and then, allied with European groups, in the European Union. The pressure eventually led the European Commission to ask a scientific committee to investigate animal welfare issues on farms. The committee agreed that the standard cages were incompatible with an acceptable level of welfare for laying hens and recommended a ban. That recommendation was accepted by the governing body of the EU in 1999 but, to ensure that producers would have plenty of time to phase out the equipment in which they had invested, its implementation was delayed until 2012. After that date, although hens could still be kept in cages, they had to have a little more space—a minimum of 750 square centimeters—plus access to a nest box, some litter material to allow pecking and scratching, and enough perches for all the birds in the cage.[60]

The nest box was important because even hens who have never known a situation in which they could lay their eggs in a nest have a strong instinct to do so. Konrad Lorenz described the laying process as the worst torture to which a battery hen is exposed:

> For the person who knows something about animals it is truly heart-rending to watch how a chicken tries again and again to crawl beneath her fellow cagemates, to search there in vain for cover. Under these circumstances hens will undoubtedly hold back their eggs for as long as possible. Their instinctive reluctance to lay eggs amidst the crowd of their cagemates is certainly as great as the one of civilized people to defecate in an analogous situation.[61]

Lorenz's view has been supported by a study in which hens were able to gain access to a nesting box only by overcoming increasingly difficult obstacles. Their high motivation to lay in a nest was shown by the fact that they worked just as hard to reach the nesting box as they did to reach food after they had been deprived of it for twenty hours.[62] Hens have evolved an instinct to lay eggs in privacy to protect their eggs and, subsequently, chicks, but another possible reason is that the vent area becomes red and moist when the egg is laid, and if this is visible to other birds, they may peck at it. If this pecking draws blood, further pecking will result, which, as we've seen, can lead to cannibalism.

Hens also provide another kind of evidence that they never lose their nesting instinct. Some of my friends have adopted hens who were at the end of their commercial laying period and about to be sent to the slaughterhouse. When these birds are released in a backyard and provided with some straw, they immediately start to build nests—even after more than a year spent in a bare metal cage. Swiss scientists have even investigated what kind of litter hens prefer and found that both caged hens and hens who had been reared on litter preferred oat husks or wheat straw; as soon as they discovered that they had a choice, none laid eggs on wire floors or even on synthetic grass. Significantly, the study found that while nearly all the hens reared on litter had left the nesting boxes forty-five minutes after they were admitted to them, the cage-reared birds seemed to be so entranced with their newfound comforts that at the end of this period 87 percent of them were still sitting there![63]

This story is repeated with other basic instincts thwarted by the cage system. Two scientists watched hens who had, for the first six months of their lives, been kept in cages so small that they could not flap their wings. Within the first ten minutes after the hens were released from the cages, half of them had already flapped their wings.[64] The same is true of dustbathing—an instinctive activity that has been shown to be necessary for maintaining feather quality. A farmyard hen will find a suitable area of fine soil and then form a hollow in it, fluffing up the soil into her feathers and then shaking energetically to remove the dust. The need to do this is instinctive, and present even in caged birds. One study found that birds kept on wire floors had "a higher denudation of the belly" and suggested that "the lack of appropriate material for dustbathing may be an important factor, as it is well known that hens perform dustbathing activities directly on the wire floor." Indeed, another researcher found that hens kept on wire actually engage in dustbathing-like behavior—without any dust to fluff into their feathers—more often than birds kept on sand, although for shorter periods of time. The urge to dustbathe is so strong that hens keep trying to do so, despite the wire floors, and rub the feathers off their bellies in the process. Again, if released from the cages, these birds will take up dustbathing with real relish. It is wonderful to see how a dejected, timid, almost featherless hen can, in a relatively short period, recover both her feathers and her natural dignity when put into a suitable environment.[65]

To appreciate the constant and acute frustration of the lives of caged hens in modern egg factories it is best to watch a cage full of hens for a short period. They seem unable to stand or perch comfortably. Even if one or two birds were content with their positions, so long as other birds in the cage are moving, they must move too. It is like watching three people trying to spend a comfortable night in a single bed—except that the hens are condemned to this fruitless struggle for an entire year rather than a single night. Finally, in most cages there is one bird—maybe more than one in larger cages—who has lost the will to resist being shoved aside and pushed underfoot by other birds. Perhaps these are the birds who, in a normal farmyard, would be low in the pecking order; but under normal conditions this would not matter so much. In the cage, however, these birds can do nothing but huddle in a corner, usually near the bottom of the sloping floor, where their fellow inmates trample over them as they try to get to the food or water troughs.

Europe's enriched cages are a significant improvement over the smaller cages that were used previously in Europe and are still widely used in the United States. Nevertheless, the enriched cages still confine hens to a space that is far too small to meet their needs. For that reason, animal advocates have pressed for cage-free conditions for all hens. In Germany, by 2021 the percentage of hens in cages had fallen to 6 percent, and federal and state governments have agreed to phase them out by 2025.[66] Also in 2021, after Compassion in World Farming collected 1.4 million signatures for a European Citizens Initiative to "End the Cage Age"—the most signatures ever obtained for an initiative under this mechanism for making a formal petition—the European Commission held hearings, examined alternatives to cages, and finally announced that by 2027 it would ban all cages for laying hens.[67]

In the United States progress lags behind Europe. In 2022, 65 percent of laying hens were still in standard cages. According to guidelines set by United Egg Producers, an industry association, eggs can be labelled "UEP Certified" if the cages allow hens 67 square inches of space, or 432 square centimeters.[68] (A standard sheet of American letter paper measures 8.5 by 11 inches, or 93.5 square inches.) To see how utterly inadequate this is, we can look at the findings of a study conducted at the Houghton Poultry Research Station, in Britain, on the space required by hens for various activities. The researchers found that the typical hen at rest physically occupies an area of 637 square centimeters, which is already almost 50 percent more than the UEP guidelines require. The UEP guidelines, therefore, are compatible with the hens being so jammed into cages that they are constantly pressing against each other. But that's not all; the researchers also found that if a bird is to be able to turn around at ease, she would need a space of 1,681 square centimeters, or almost four times the amount allowed by the UEP guidelines.[69]

In addition to allowing hens to be crammed together, the UEP guidelines allow cages to be completely barren, consisting of bare wire floors and walls, with no nesting boxes, perches, or litter—and that is how the majority of American laying hens still live. Under these conditions every natural instinct the birds have, except the desire to eat and drink, is frustrated. The hens cannot turn around with ease, cannot walk around, scratch and peck at the ground, dustbathe, build nests, perch, or stretch their wings. They cannot keep out of each other's way, and weaker birds have no escape from the attacks of stronger ones, already maddened by the unnatural conditions. For a hen, this is a miserable life, and as you read these words, there are hundreds of millions of hens living like this in the United States alone, and many more in other countries that have not banned the standard cage for laying hens. Given a choice, hens familiar with both grassed runs and cages will go to the run. In fact, most of them will prefer a run with no food on it to a cage that does have food in it.[70]

Hens can live seven or eight years, but after a year or so of laying, they start to lay fewer eggs. Commercial producers therefore get rid of them after twelve to eighteen months and bring in a new batch. The "spent hens"—as the U.S. industry refers to these birds, who are still laying eggs, just not quite enough to maximize returns for the producers—have little commercial value. As a Poultry Site article puts it, "The commercial value of spent hens has long been considered negligible," and so the animals may be rendered into animal feed, pet food, or "simply composted or just buried after being euthanized because of their low market value."[71] To refer to this as "being euthanized" suggests that the birds are getting euthanasia, which literally means "a good death," but that is a grossly misleading description of the methods used to bring mass death to tens of thousands of birds. A report from the European Food Safety Authority, using a less offensive term for the birds, says that "For end-of-lay hens, slaughter is not always an option" either because of "the costs and work involved in catching, crating, transporting and slaughtering birds with a very limited economic value," or "to spare the birds the stress of being caught (depending on the method), crated and transported." If the birds are killed on the property, the report lists various acceptable methods of killing, including gassing the entire shed full of birds.[72]

As we have already seen with deformed and ill chickens raised for meat, when animals have no commercial value, there are always producers who get rid of them in whatever way is quickest and cheapest. So we should not be too surprised that in 2003 workers for an egg producer in San Diego, California, were seen by a neighbor emptying buckets full of live, squirming hens into a wood chipper. The county Animal Services Department investigated, and the farm owners acknowledged that they had disposed of 30,000 hens by that method. They said, however, that they were "just following professional advice" from two veterinarians. Neither they, nor the veterinarians, were prosecuted. Two years later, another egg producer, this time in Missouri, was found to have gotten rid of thousands of live hens by throwing them in a dumpster.[73] Very probably, similar occurrences happen far more commonly than we know about but go undetected.

In contrast to the condemned prisoner who gets a special meal before being executed, the condemned hens are likely to get no food at all. According to United Egg Producers guidelines, "Catching and transport must be planned so that feed is withdrawn no more than twenty-four hours prior to slaughter or depopulation." A European Union guide to poultry transport also cites a maximum time of twenty-four hours for hens to be without food but suggests the point of this is to minimize feces production.[74] That can't be a factor, though, when the hens are being "depopulated" and composted, as they often are in the United States. Perhaps the real reason for not feeding the hens was given by an article in *Poultry Tribune* back in 1974, when criticism of factory farming was rare and producers were not as careful about their public image as they are today: "Take feed away from spent hens" advised the headline above an article telling farmers that food given to hens in the thirty hours prior to slaughter is wasted, since they won't get paid any more for food that remains in the digestive tract.[75] In any case the readiness with which birds are deprived of food for twenty-four hours before they are killed is just one more indication of the producers' indifference to the suffering of birds who are powerless to protest what is being done to them.

I began this chapter with the "depopulation" of around a million pigs during the COVID-19 pandemic. But a much larger number of chickens, turkeys, and ducks periodically suffer the same fate, not because slaughterhouses lack the capacity to kill them but because there are periodic outbreaks of avian influenza, popularly known as "bird flu." When a flock is affected, the birds cannot be taken to slaughter, as that might spread the disease. The most common response, therefore, is to shut off the ventilation in the shed and add heaters, which means that the birds will die, slowly, of heatstroke. Another technique is "ventilator shutdown alone," without using heaters. Without ventilation, eventually the body heat of the animals will cause death by heatstroke, but it will take longer and more birds will survive, possibly to be buried alive with the corpses that are being composted. In 2014–2015, more than 50 million chickens and turkeys in the United States either died of avian influenza or had to be killed because some birds in large flocks had been infected.[76] Rembrandt Enterprises, an egg producer near Storm Lake, Iowa, killed all of its 5.6 million hens in that time. In 2022, however, another outbreak of the virus occurred in the United States. It infected birds at Rembrandt Enterprises again, and once more they killed their entire flock, this time 5.3 million hens, using the same method used to kill the pigs: shutting down the ventilation and adding additional heat, a method accepted by the Iowa Department of Agriculture and Land Stewardship.[77] Several other producers were also affected and responded similarly. According to USDA figures, in 2022 more than 50 million birds were killed, by the use of ventilator shutdown alone, or by "VSD +"—a term that stands for ventilator shutdown combined with additional heat, and possibly in some cases combined with other, unspecified, methods.[78]

If you are in any doubt about whether VSD+ is an inhumane way to kill birds, you can watch videos of hens dying in this manner. The videos were taken as part of research conducted at North Carolina State University, paid for by the U.S. Poultry and Egg Association and obtained through public records requests by Animal Outlook, an animal advocacy organization.[79] They show chickens dying slowly and in evident distress as the heat rises and they have difficulty breathing. Yet instead of developing quicker methods of killing the birds when the next outbreak of avian influenza should appear, the industry appears to have decided that VSD+ is acceptable in such situations.

Many veterinarians consider the acceptance and implementation of such methods by members of their profession to be a deeply disturbing violation of veterinary ethics. In response to such objections, Cia Johnson, the head of the animal welfare division of the American Veterinary Medical Association (AVMA), appeared to be lobbying veterinarians working with farm animals to support VSD+ when she told a conference in 2022, "Some of these methods are at risk of leaving the guidelines, I think you probably have an idea of what those methods might be. We need data to support them staying in the document."[80] A petition was circulated calling on the AVMA to reclassify "heatstroke-based depopulation methods" as "not recommended," pending a further inquiry. The petition was signed by 278 AVMA members, but an AVMA subcommittee refused to allow the resolution that the petition was requesting to be put to a vote.[81]

Avian influenza was also present in Europe in 2022. The European Food Safety Authority, in an 83 page "Scientific Opinion" states flatly: "Ventilation shutdown should not be used as a killing method," and describes other more humane methods of killing birds when they cannot be sent for slaughter.[82] Nevertheless, in France, at least 14 million birds were suffocated. One farmer said that they had to finish off the birds with a shovel.[83] Other producers used carbon dioxide, which is at least quicker than shutting down the ventilation, with or without additional heat.

PIGS

WHEN GEORGE ORWELL CHOSE PIGS TO BE THE LEADERS OF THE revolution of farm animals against humans in his novel *Animal Farm*, his choice was defensible on scientific as well as literary grounds. Of all the animals commonly eaten in the Western world, pigs appear to be the most intelligent. Their problem-solving abilities are comparable to, and perhaps superior to, those of dogs. Although it is the capacity to suffer or enjoy life that entitles all sentient beings to equal consideration of their interests in avoiding pain and experiencing pleasure, the high intelligence of pigs must be borne in mind when we consider whether the conditions in which they are reared do provide them with what they need to enjoy their lives. Animals of different capacities have different needs. Common to all is a need for physical comfort and the avoidance of pain, extreme heat and cold, hunger and thirst. Pigs also need stimulation. Without it, they suffer from boredom.

Researchers at Edinburgh University studied commercial pigs released into a semi-natural enclosure and found that in those environs, the pigs form stable social groups, build communal nests, and keep the nest clean by using dunging areas well away from it. They remain active, spending much of the day rooting around the edge of the woodlands. Pigs are social, and when allowed to be free-ranging, will form groups that consist of about three sows and their offspring. When pregnant sows are ready to give birth, they leave the communal nest and build their own nest, finding a suitable site, scraping a hole, and lining it with grass and twigs. There they give birth and live for about nine days, until they and their piglets rejoin the group.[84]

Pigs in today's intensive production units cannot follow these innate behavior patterns. They have nothing to do but eat, sleep, stand up, and lie down. They have no straw or other bedding material, because it requires employees to remove the soiled straw and replace it with clean bedding. Pigs kept in this way can hardly fail to put on weight, but in contrast to the pigs observed by the Edinburgh researchers, these pigs will be bored and unhappy. Scientific studies have shown that pigs kept in a barren environment are so bored that if they are given both food and an earth-filled trough, they will root around in the earth before eating.[85] When kept in barren, overcrowded conditions, pigs are prone to "vice," as hens are. Instead of feather-pecking, pigs take to biting each other's tails. This leads to fighting in the pig pen and reduces gains in weight. Since pigs do not have beaks, farmers cannot debeak them to prevent this, but they have found another way of eliminating the symptoms without altering the conditions that cause the trouble: They cut off the pigs' tails. This is a routine procedure on commercial pig farms in the United States and some other countries. As one review of the procedure states:

> Methods of tail docking include cutting with a knife or scalpel, cutting with a hot docking iron, or application of a constrictive rubber ring. All methods are commonly performed without analgesia or anesthesia, and all likely result in some degree of pain.[86]

Why is this necessary, when pigs on more traditional farms do not need to have their tails cut off? According to Professor David Fraser, of the University of British Columbia's Animal Welfare Program:

> The probable underlying cause . . . is that pigs are using species-typical activities in an unusual way because no suitable object is available. The lower incidence of tail-biting in units with straw bedding is probably due, at least in part, to the "recreational" effects of the straw.[87]

The problem seems to be that bored pigs gnaw at any attractive object, and if gnawing on the tail of another pig should produce an injury and draw blood, other pigs will be attracted to the blood and begin biting in earnest.[88] A carefully controlled Danish study has shown that a combination of reduced stocking density and the provision of straw is just as effective in reducing tail biting as tail docking. These conditions are used in Sweden, where tail docking is banned.[89] In fact, in contrast to the United States, tail docking as a routine procedure is banned across the entire European Union. The law is often evaded, however, because any incident of tail biting can permit producers to cut off the tails of all their pigs. As cutting off tails is cheaper than reducing stocking densities and providing straw, the pigs' tails are likely to go. In France, although tail docking has been illegal since 2003, the law was ignored until finally, in 2022, the efforts of the animal organization L214 led to the first case in which a producer was fined for routinely cutting off the tails of its pigs.[90]

The pig industry was slower to move to total confinement than the chicken and egg industries, but in the United States today almost all the small family farmers with fifty to a hundred pigs have been pushed out of business by huge pig factories. The U.S. 2017 agriculture census showed that 93.5 percent of pigs sold in that year came from producers with more than 5,000 pigs.[91] Tyson Foods, a giant of the chicken industry, has been expanding into pig production and in 2021 sent 469,000 pigs to slaughter every week.[92] In these pig factories, pigs are born and suckled in a farrowing unit, raised initially in a nursery, then grown, and finally "finished." They are sent to market at between six and seven months of age, weighing close to 280 pounds.[93] The desire to cut labor costs is a prime reason for the shift to confinement. Automated feeding means that very little labor is required. Another consideration is that pigs with less room to move burn up less of their food in "useless" exercise and so can be expected to put on more weight for each pound of food consumed. In all of this, as one pig producer said, "What we are really trying to do is modify the animal's environment for maximum profit."[94]

In addition to stress, boredom, and crowding, modern pig confinement units create physical problems for the pigs. One is the atmosphere, which is high in ammonia from the feces and urine of the pigs. The herdsman at Lehmann Bros. Farms, of Strawn, Illinois, has described it like this:

> The ammonia really chews up the animals' lungs. . . . The bad air's a problem. After I've been working in here a while, I can feel it in my own lungs. But at least I get out

of here at night. The pigs don't, so we have to keep them on tetracycline [an antibiotic], which really does help control the problem.[95]

This is not a producer with unusually low standards. The year before this statement was made, Lehmann Bros. had been named an Illinois Pork All-American by the National Pork Producers Council.

The ammonia also pollutes both air and water. A survey of 130 lakes in Iowa, the country's largest producer of pigs, showed that a reduction of only 10 percent in emissions from pig farms would lead to an increase in residential property values of $80 to $400 million (in 2014 dollars), with about 10,000 more recreational visits to lakes every year and two to three fewer deaths of infants. (That's human infants; the impact of the ammonia on the pigs was not included in this survey.)[96]

Perhaps the most alarming aspect of intensive pig farming, however, is the treatment of the sows—the mothers of the pigs sold to be eaten. "The breeding sow should be thought of, and treated as, a valuable piece of machinery whose function is to pump out baby pigs like a sausage machine."[97] So said a leading corporate manager with Wall's Meat Company at the time when the intensification of pig production was emerging. The United States Department of Agriculture also encouraged producers to think of the sow as "a pig manufacturing unit."[98]

Under the best conditions there is little joy in an existence that consists of pregnancy, birth, having one's babies taken away, and becoming pregnant again so that the cycle can be repeated—and sows do not live under the best of conditions. In the United States, 75 percent of them are kept in gestation stalls—individual metal stalls two feet wide and six feet long, or scarcely bigger than the sow herself; or they may be chained by a collar around the neck. As the animal scientist Temple Grandin put it, "Basically you're asking a sow to live in an airline seat."[99] They will live like this for the duration of their pregnancies, usually around 114 days. During all that time, they will be unable to walk more than a single step forward or backward, or to turn around, or to exercise in any other way.

Sows kept in stalls show many signs of stress, such as gnawing the bars of their stalls, chewing when there is nothing to chew, waving their heads back and forth, and so on. This is known as stereotypical behavior. Anyone who has been to an old-style zoo that keeps lions, tigers, or bears in barren concrete enclosures will have seen stereotypical behavior—they pace endlessly up and down along the fences of their cages. The sow does not have even this opportunity.

In Europe in the 1990s, animal organizations objected strongly to keeping sows in individual stalls. The European Union asked its Scientific Veterinary Committee (SVC) to assess the welfare of sows in stalls. After receiving a damning report, which concluded that sows in stalls "may well be depressed in the clinical sense,"[100] the European Commission decided that sow stalls are incompatible with acceptable standards of animal welfare and that alternative and better systems of keeping sows are feasible and involve little or no extra cost. The Commission therefore recommended a period during which sow stalls were to be phased out. This was completed in 2013, and since then sow stalls have been illegal across the entire European Union, except for the first four weeks of pregnancy. Instead, in Europe sows are now typically kept in groups, where they are more active, able to move around, and also have access to straw, hay, soil, or other materials they can manipulate or root around in with their snouts, so the stereotypical behavior that is typical of sows in stalls is rare.

Sows in the United States, unfortunately, are still mostly kept in the conditions that Europe rejected twenty years ago. Nine states, including California, Michigan, and Massachusetts, have outlawed sow stalls, but these states account for only 3 percent of U.S. pig production. A ban in Ohio will come into effect in 2026, but the Economic Research Service of the U.S. Department of Agriculture expects that even after that, fewer than 10 percent of pigs will be in states with laws that require producers to give their pigs room to turn around.[101]

When the sow is ready to give birth she is moved—but only to a "farrowing pen." (Humans give birth, but sows "farrow"—another word that is designed to separate us from animals.) Here the sow may be even more tightly restricted in her movements than she was in her stall, constrained by an iron frame that keeps her in a prone position. The ostensible purpose is to stop the sow rolling onto and crushing her piglets, a problem that has been exacerbated by the deliberate selective breeding of sows who will give birth to a large number of piglets. The problem could therefore be reduced by abandoning this kind of genetic selection, or by providing sows with more space and straw or other materials they could use to build a nest. The U.S. Department of Agriculture acknowledged long ago that "the sow kept in a crate cannot fulfill her strong instinct to build a nest" and that this frustration could contribute to problems in giving birth and nursing her infants.[102] A sow who is confined both while pregnant and while nursing is tightly restricted for almost the whole of her life, in a monotonous environment that she cannot alter.

In addition to all of the misery just described, sows and boars, like the breeding birds used to produce chickens raised for meat, are kept permanently hungry. The animals being fattened for market are genetically selected to have big appetites and grow quickly, but to give breeding animals more than the bare minimum required to keep them reproducing will make them obese;

and anyway this is, from the producer's point of view, simply a waste of money. A study showed that pigs fed the rations recommended by the Agricultural Research Council in Great Britain are getting only 60 percent of what they would eat if they had more food available. Moreover, their readiness to press levers in order to get additional food was much the same after eating their daily rations as it was before, indicating that they were still hungry immediately after feeding. As the scientists concluded:

> Commercial levels of feeding for pregnant sows and boars, whilst meeting the needs of the producer, do not satisfy feeding motivation. It has often been assumed that high production levels cannot be achieved in the absence of adequate welfare. Yet the hunger resulting from the low food levels offered to the pig breeding population may act as a major source of stress.[103]

The European Commission's Scientific Veterinary Committee reached a similar conclusion: "The food provided for dry sows [sows not giving milk to their piglets] is usually much less than that which sows would choose to consume, so the animals are hungry throughout much of their lives."[104] Once again, the producer's profits and the interests of the animal are in conflict. It is truly amazing how often this can be demonstrated—while the agribusiness lobby constantly assures us that only happy, well-cared-for animals can be productive.

I can't think of a better summary of the true nature of the intensive pig industry than these words from Eric Schlosser, author of the bestseller *Fast Food Nation*:

> Hogs are intelligent and sensitive creatures capable of multistage reasoning like dolphins and apes, with a social

structure similar to that of elephants. Hogs can recognize themselves in a mirror, differentiate one person from another, remember negative experiences. And they like to be clean. Their lives in hog factories scarcely resemble how they've been raised for millenniums. They arrive as small piglets, live crammed together amid one another's filth, and leave a few months later for the slaughterhouse—never having enjoyed a moment outdoors during their entire time at the shed. The foulness of these places, for the animals that live in them and the people who live near them, truly defies words.[105]

CALVES RAISED FOR VEAL

OF ALL THE FORMS OF INTENSIVE FARMING NOW PRACTICED, THE veal industry has long had the worst reputation. The essence of producing the premium product known as "white veal" is the feeding of a high-protein food to confined, anemic calves in a manner that will cause the calves to have tender, light-colored flesh that will be sold at premium prices to expensive restaurants. ("White" veal is not really white—a U.S. Department of Agriculture document described it in terms less likely to be favored by those trying to market the product as "a grayish pink color."[106]) The veal industry does not compare in size with poultry, beef, or pig production, but it is worth our attention, first to understand the extreme degree of exploitation to which an animal-producing industry is willing to go, and second to recognize (but also scrutinize) the efforts that the industry has made, since the first edition of this book, to overcome its dire reputation.

Veal is the flesh of a young calf. The term was originally reserved for calves killed before they had been weaned from their mothers. The flesh of these very young animals was paler and more tender than that of a calf who had begun to eat grass; but there was not much of it, since calves begin to eat grass when they are a few weeks old and still very small. The small amount available came from the unwanted male calves produced by the dairy industry. A day or two after being born they were trucked to market where, hungry and frightened by the strange surroundings and the absence of their mothers, they were sold for immediate delivery to the slaughterhouse.

In the 1950s veal producers in Holland found a way to keep calves alive longer without the flesh becoming red or less tender. The trick relies on keeping the calf in highly unnatural conditions. Specialist veal producers bought the unweaned calves at auction and took them straight to a confinement unit. There they were put in individual stalls or crates, too narrow for them to turn around, and not large enough for them to walk even one or two steps. Sometimes the stalls had open backs, but the calves were tethered to the front of the stall.

Under such conditions the young calves sorely miss their mothers, and also miss something to suck on. The urge to suck is strong in a baby calf, as it is in a baby human. From their first day in confinement—which may well be only the third or fourth day of their lives—they drink from a plastic bucket. In doing research for the first edition of this book I visited a veal producer who showed me his calves, their heads following us as they looked out from their narrow stalls. He told me to offer my thumb to one of the calves. With some trepidation, I did so, and instantly the calf was sucking vigorously on it, as a baby would suck on a dummy.

When the calves lie down, it would be on bare floors, without straw, hay, or other bedding. As with pigs, providing bedding would increase costs, both for the bedding itself and for the extra labor required to change it. But with veal calves, straw or hay bedding is avoided for another reason. The calves are deliberately kept low in iron, because iron causes the flesh to become red. Straw and hay contain iron, and the young calves, lacking both iron and roughage, would eat it. Similarly, they are never allowed out in a field, because they would eat grass, and it was also thought that if they were allowed to move around, they would develop muscles that would make their flesh less tender. Manuals on veal raising advised producers to check the iron levels in their water, fit an iron filter if they were high, and to keep rusty metal away from anywhere the calves could lick it and absorb iron.[107]

Thus the calves lived for about sixteen weeks, leaving their stalls only when taken out to be transported to slaughter. The benefit of the system, from the producers' point of view, was that at this age the veal calf may weigh 400 pounds, instead of the ninety pounds that newborn calves weigh. Since veal fetches a premium price, rearing veal calves in this manner is a profitable occupation. But when these practices were exposed—as they were by some animal welfare organizations, and in the first edition of this book—people started avoiding veal. Opposition was particularly strong in Europe. The European Union sought advice from its Scientific Veterinary Committee, and its report condemned the methods then in use. This led, in 1997, to a major victory for the animal movement: a European Union decision to ban individual stalls, much like they did sow stalls, and require that all calves be fed a nutritionally adequate diet. The regulations came into effect at the end of 2006, and since then calves raised for veal in Europe have been housed in groups, and fed a diet that is adequate for iron and includes some roughage.

Protests against white veal production were also building in the United States. The American Veal Association (AVA), a producers' organization, responded to the criticism—and, to their credit, not merely by trying to fool consumers into believing that veal calves enjoyed living in solitary confinement. In 2007, the AVA adopted a resolution calling on all veal producers to transition, within ten years, to group housing that would allow the calves to stand, stretch, lie down, turn around, and socialize with other calves. In January 2018, the AVA announced that this mission had been accomplished and that "all AVA-member companies and individuals involved in veal production have successfully transitioned to group housing and no tethers."[108]

These changes do give many veal calves in Europe and the United States better lives than those who lived before the changes occurred, but problems remain in both jurisdictions. The EU regulations allow calves to be kept in individual pens up to eight weeks of age, although the pens must be wide enough to allow the calves to turn around, and the AVA standards allow individual housing up to ten weeks. This is, ostensibly, because the young calves are delicate and more likely to succumb to illness, but that is largely because they are not with their mothers. The isolation also comes at a heavy cost to the calves. Significantly, in 2021 the U.S. Dairy Cattle Welfare Council recognized this when it issued a position statement supporting "social housing in pairs or groups for dairy calves starting from one to four days of age." Social housing has, the council said, been shown to improve cognitive development, social skills and play, emotional support during stressful procedures, to decrease stress at weaning, and to aid in reducing anxiety and stress associated with separation from the mother at birth. Although the council was addressing the raising of female calves, who would become dairy cows, there is no reason to think that the benefits of social housing would be less for the male offspring of the same cows.

Then there is the issue of anemia. Although the AVA claims that the calves raised by its members are not anemic,[109] Dr. James Reynolds, a professor of veterinary medicine at the Western

University of Health Sciences, in Pomona, California, believes that "because they want veal to be soft and pale, they keep [the calves] anemic." He acknowledges that because it would be unethical to create disease, and anemia is a disease, AVA producers monitor the calves' iron levels and give the calves iron injections if their levels are too low; nevertheless, he says, they are on the borderline of anemia.[110]

The AVA can require its own members to produce veal in accordance with its standards, but not all U.S. veal producers are AVA members. Nine states have laws requiring that calves have room to move around and carry out natural behaviors, like grooming themselves: Arizona, California, Colorado, Kentucky, Maine, Massachusetts, Michigan, Ohio, and Rhode Island (though the laws do not prohibit social isolation, because they allow individual stalls or pens, as long as they are large enough). But, as with the laws banning sow stalls that prevent the sows turning around, this list does not include the biggest producers—in this case, Wisconsin and Indiana; nor Pennsylvania, which has Mennonite and Amish communities that often raise veal calves. Professor Reynolds says that farmers who are members of these communities tend to use the methods introduced into the United States from Holland in the 1950s and 1960s that are now rejected by the AVA.

The good news is that in the United States, veal consumption has fallen sharply from around eight pounds per person per year in the 1940s to less than one pound in 1995 and to only one fifth of a pound in 2020. Italian restaurants now make up a large share of the U.S. veal market. Price is a factor in this low level of consumption—veal is significantly more expensive than beef—but so is widespread awareness of the animal welfare issues I have summarized here. In the European Union, where it is illegal to raise veal calves by methods still used in the United States, veal consumption has not fallen to the same extent.[111]

Separating calves from their mothers immediately after birth is cruel, and isolating them for weeks afterward makes it worse. Keeping the flesh pale as the calves pass the age at which they would be eating grass deprives the calves of iron and roughage and is a health risk. Moreover, all of these aspects of "white" veal production serve no other purpose than providing a luxury product for people who put the satisfaction of their gourmet tastes above the suffering of the animals they are eating.

COWS KEPT TO PRODUCE MILK

AS WE HAVE SEEN, THE VEAL INDUSTRY IS AN OFFSHOOT OF DAIRYing. Dairy producers must ensure that their cows become pregnant every year, for otherwise their milk will dry up. Their babies are taken from them at birth, an experience that is as painful for the mother as it is terrifying for the calf. The mother often makes her feelings plain by constant calling and bellowing for her calf—and this may continue for several days after her infant calf is taken away. Some female calves will be reared on milk substitutes to become replacements of dairy cows when they reach the age of milk production, at around two years. Some others will be sold at between one to two weeks of age to be reared for beef in fattening pens or feedlots. The remainder will be sold to veal producers.

The bucolic picture of the dairy cow playing with her calf in the pasture is no part of commercial milk production. Today the typical dairy cow is no longer peacefully munching grass in a field. Instead, she is likely to be a fine-tuned milk machine, reared in a zero-grazing dairy facility. She will have been bred to produce enormous amounts of milk, up to sixty liters (fifteen U.S. gallons) per day at peak production.[112] Some are kept in individual pens with only enough room to stand up and lie down; others are kept in stanchions in which they cannot even adopt the majority of prone positions that cows can be observed to use. More fortunate ones have access to a small outdoor yard, and only the really lucky ones are free to graze in a meadow. Her environment is completely controlled: She is fed calculated amounts of feed, temperatures are adjusted to maximize milk yield, and lighting is artificially set, usually with sixteen hours of light and only eight hours of darkness, as that has been found to induce higher milk production.[113]

After her first calf is taken away, the cow's production cycle begins. She is milked twice, sometimes three times a day, for ten months. After the third month she will be made pregnant again. She will be milked until about six or eight weeks before her next calf is due, and then again as soon as the calf is removed. Although cows can live for fifteen to twenty years, this intense cycle of pregnancy and hyperlactation can last only about five years before her milk production declines and she is considered "spent."

Then, often lame or suffering from one or more painful medical conditions, she will be sent to an auction ring or other market, and then to one of the slaughterhouses that accept cows from dairy farms. Only a few slaughterhouses have the specialized equipment required to handle these large animals, so she may be trucked across the country for more than one day. Bred to produce a lot of milk, suddenly she is no longer being milked. Her udders may become engorged and painful. For her owners, she is of little value; her flesh is good for only hamburger or dog food, and her welfare no longer matters.

Dairy cows are sensitive animals who manifest both psychological and physiological disturbances as a result of stress. They have a strong need to identify with their "caretakers." On traditional dairy farms, that would be the person who spends a lot of time with them, knows them each by name, milks them, feeds them, and is familiar with the distinctive personality of each animal. Today's system of dairy production does not allow the workers more than five minutes a day with each animal. The majority of dairy farms still had fewer than 200 cows in 1988. Today most dairy cows are controlled by producers with more than 1,000 cows, and a third are on establishments that have more than 2,500 cows.[114] In such large facilities, there is little opportunity for personal interaction between employees and cows.

During the first forty years after the first edition of this book, no one contradicted my statement that all commercial milk producers make their cows pregnant every year and then take the calf away so that the milk can be sold. Then I heard about How Now Dairy, started by Les Sandle, a third-generation Australian dairy farmer, and his partner, Cathy Palmer, who describes herself as "a passionate animal rights activist." Sandle had been complaining about the direction in which the Australian dairy industry was going—toward mega-dairies, as in the United States—and in 2012 he and Palmer decided to do something about it. Their goal was to create an ethical dairy farm based on the radical belief that ethics and compassion do not have to clash with economics. Their sixty-four-acre farm, about 200 kilometers north of Melbourne, has cows grazing in paddocks dotted with trees, but its most remarkable feature is that the calves are allowed to stay with their mothers and, at least until they are weaned at about four months of age, they can drink as much of their mothers' milk as they wish. According to How Now Dairy, the calves drink about four or five liters a day, and never more than eight liters, but as their mother produces about twenty liters a day, that still leaves plenty to sell. As commercial milk producers go, How Now is tiny, but it has, at the time of writing, produced over 700,000 liters of milk without separating the calves from their mothers or killing a single calf.

When I first heard that How Now does not kill any calves, my reaction was: "That's impossible! They will be overrun with nonproductive male calves who will soon turn into bulls!" But here modern technology comes to the rescue of what is otherwise a very natural way of farming. There are no male calves, because How Now uses artificial insemination with sexed semen to ensure that only female calves are born. The females all grow up to become part of the herd. There aren't as many calves as most dairy farms would have, because the cows only have a calf every eighteen to twenty-four months. That means that they continue to be productive for longer, if at a lower level, and, as How Now puts it, these animals "have a quality of life and do not just exist as milk machines."

Sandle and Palmer aren't doing this just because it makes them feel good about the way they treat their cows. They set out to challenge the idea that if you are going to produce milk, you have to do bad things to cows and their calves. How Now's milk is more expensive than milk produced by the standard methods, but their business has prospered, thanks to customers willing to pay a premium for milk from a dairy farm that is run on the cows' terms.[115]

There are a few such dairy farms in other countries as well. Ahimsa Dairy Foundation, in Rutland, England, allows the calves to stay with their mothers. In Germany, a dairy farm run by Anja Hradetsky, who also allows her calves to stay with their mothers, was featured in a documentary that describes this form of dairy farming as "a niche within a niche"—the first niche being organic dairy farms, which constitute 3.5 percent of all German dairy farms; and the second being dairy farms that leave the calves with their mothers, which are just 2 percent of organic dairy farms. If you do the math, that works out to less than one dairy cow in a thousand in Germany not being separated from her calf soon after giving birth.[116] The care that dairy farmers like Sandle and Palmer, Hradetsky, and the founders of Ahimsa Dairy Foundation show for their animals is admirable; but that makes it all the more regrettable that there is no sign of this approach having any impact on milk production as a whole.

COWS RAISED FOR MEAT

ANIMALS RAISED AND KILLED FOR BEEF ARE COMMONLY REferred to as "cattle." That word, a variant of "chattel," originally referred to movable personal property. To avoid the suggestion that the animals raised for beef are, by their nature, items of human property, and instead encourage you to consider them as individual animals, I will refer to them, whether male or female, as "cows," using the word in a manner that, as the *Encyclopedia Britannica* states, is already used "in common parlance" to refer to "a domestic bovine, regardless of sex and age, usually of the species Bos taurus."[117]

Traditionally, cows raised for beef in America have roamed freely over the vast open spaces that we see in cowboy movies. But as a supposedly humorous article in the *Peoria Journal Star* indicated, the modern range isn't what it used to be:

> A cowboy's home ain't necessarily on the range. More 'n likely, home is a feedlot where the closest a beef comes to the smell of sage is in a pot roast. This is cowboy'n' modern-like. This is Norris Farms, where instead of

> running 700 head on 20,000 acres of sparse-grass prairie, they run 7,000 head on 11 acres of concrete.[118]

By comparison with chickens, pigs, veal calves, and dairy cows, beef cattle still see more of the "the range," but the time they have there has been diminished. In the days of the cowboy era, cows would have roamed freely for about two years. Now, most will spend only six to eight months on grass and will then be trucked long distances to a feedlot to be "finished"—that is, to be fattened to market weight and condition on a richer diet that includes corn and other grains. Then, at about fourteen months of age, they are sent for slaughter.[119]

The shift to feedlot fattening of cows has been the dominant trend in the beef industry for several decades. Outdoor feedlots are now used to fatten three of every four cows raised for beef. Within that trend, there has also been a move toward large feedlots, with four in every ten cows fed in the United States now coming from feedlots with more than 32,000 animals; some feedlots hold 100,000.[120] They are profitable because cows fatten more quickly on grain than on grass, and the grain produces beef marbled with fat, which Americans prefer; and besides, in the United States corn is heavily subsidized and sells for less than the cost of production.

But the standard diet for these animals ignores the fact that their stomachs have evolved to digest grass. When most of their diet is grain, the rumen—the part of their stomach where digestion begins and the food is fermented—becomes too acidic. The animal will suffer from bloat and, more seriously, the wall of the rumen may be damaged, allowing bacteria to pass into the bloodstream and contaminate the liver, which may cause abscesses. The U.S. National Beef Council Quality Audit for 2016 showed that nearly 18 percent of beef cow livers were condemned because of abscesses, and another 10 percent condemned because they were contaminated.[121]

Feedlots do not confine cows as severely as factory farms confine chickens, pigs, or veal calves, but that does not make them a suitable place for cows. Each animal can move around the compound, which may be an acre in area. Unlike veal calves, these social mammals are not isolated. Boredom from the barren, unchanging environment is more of a problem than restriction of movement.

Another very serious problem is exposure to the elements. In the United States cows are generally not provided with shade, even in states like Texas that have extremely hot summers, despite the fact that cows move into any available shade in hot weather. A 2020 review of the impact of shade on beef cows acknowledges that "several heat waves in the recent past have caused extensive death loss in feedlots."[122] Why then do U.S. feedlot owners in regions with hot summers not provide shade to their animals (as, for example, feedlot owners in northern Australia do)? Once again, economic thinking trumps animal welfare concerns. In a carefully controlled study, researchers from the Department of Animal Science and Food Technology at Texas Tech University divided cows into two groups, one with shade available and the other without. During the hot part of the day, cows in the first group used the shade, following it as the sun moved. Interestingly, the researchers noticed that cows in the second group had many more aggressive encounters with each other than those in the first group—suggesting that, like humans, cows get irritable when they are too hot. The punchline, though, is that "in west Texas, shade is generally not used in commercial feedlots because it is not thought to be cost effective."[123] The 2020 review reached a similar conclusion: "There is not a cohesive belief within the beef industry that the benefits of shade outweigh the economic investment." As a result of this narrowly economic way of thinking about cows, during a spell of hot weather in June 2022, more than 2,000 cows died from heatstroke in Kansas alone. (An NBC report of the event merely said the deaths, coming on top of higher grain prices, "added pain to the U.S. cattle industry.")[124]

In Europe and some areas of the United States, with higher rainfall totals, such as Iowa and Illinois, some beef producers have followed the lead of the poultry, pig, and veal industries and brought their animals indoors.[125] That protects animals from the weather, but always at the cost of much more crowding, since the beef producer wants the greatest possible return on the capital invested in the building. Beef cows confined indoors are generally kept together in groups, in pens rather than in single stalls. Slatted floors are often used for ease of cleaning, although beef cows, like pigs and veal calves, are uncomfortable on slats and can become lame.

FISH

INTENSIVE FISH PRODUCTION CONFINES MORE VERTEBRATE ANImals than everything else described in this chapter. When I use the term "fish" I am referring to vertebrates, sometimes called finfish, to distinguish them from shellfish like oysters and shrimps and—although they have no shells—octopuses. As we saw in chapter 1, the evidence that fish are sentient beings is strong. A scientific panel set up by the European Food Safety Authority to report on whether fish can suffer found evidence of a capacity to experience pain, fear, and distress and concluded that "the concept of welfare is the same for all the animals, i.e., mammals, birds and fish, used for human food."[126]

The evidence we have is certainly strong enough to give rise to great concern about the suffering of farmed fish, especially considering that the best recent ballpark figure for the number of farmed fish is around 124 billion.[127] That number does not include the much larger number of wild-caught fish who are made into feed for intensively farmed fish and crustaceans, estimated to be somewhere in the range of 500 billion to 1 trillion.

As these figures suggest, the intensive production of fish shares with other forms of intensive animal production the basic problem that we will look at in more detail in chapter 4: When you confine animals so that they cannot find food themselves, you have to produce food for them, and in every case, you have to put more food into the animals than you will get out of them. This still pays, commercially, if you catch fish with low commercial value and grind them up into fishmeal or fish oil, which is then sold to intensive fish producers who will feed it to fish for whom people will pay a high price. (Fishmeal is also fed to factory-farmed chickens and pigs, and some goes into pet food.) Atlantic salmon is one of the most popular farmed fish, especially in affluent countries. Salmon are carnivorous, and on average, one 4 kg salmon requires 147 fish to be killed as its food source. Farmed European eels are even worse, because they weigh only about 1 kg, and yet 79 fish are killed to provide food for them. Nile tilapia, another fish farmed in huge numbers, are largely herbivorous, but seven fish will still be killed to produce a 1 kg fish. Carp does best in this respect, for they can live on plants and algae, and a 1 kg carp is likely to have consumed only one fish.[128]

As with most intensive animal production, when economics and animal welfare point in different directions, economics wins. Fish, like other intensively farmed animals, are genetically selected for growth that is much faster than wild fish experience, and this causes abnormalities. Instead of the leg problems that chickens have, some fish may have difficulty breathing and eating.

Stocking fish more densely will result in poor water quality and higher mortality, but it can still yield a bigger profit for the producer. For some species, fish producers expect 70 to 80 percent mortality among the fry. In 2019 a Hong Kong news service described producers welcoming the prospect of a new technique for identifying parasites that, they hoped, could enable half of the young fish to survive.[129]

Different species of fish have different welfare needs. Some are social, and not stressed by having many other fish in close proximity; but others, including salmon, live largely solitary lives. In intensive confinement, these large fish, up to 75 cm long, are forced to crowd together so closely that each fish has the space equivalent of a bathtub of water. That means more stress, more aggressive behavior, damage to fins, and less oxygen in the water.

Salmon are famous for migrating across oceans and then returning to the stream in which they were born, yet at any one time, more than a billion salmon are confined in nets that thwart that instinct.[130] We don't know how much distress that causes them, but it may be responsible for the stereotypic behavior that is observed in fish farming, including swimming in endless circles inside their cages. Nile tilapia typically live, as their name suggests, in rivers. They are territorial, and the males make circular nests on the river bottom, moving gravel with their mouths. Courtship and mating then occurs within the nest area. That behavior is not possible in a cage or net suspended in the water.

As intensively produced fish reach the size at which they will be killed, they are first starved, in some cases for as long as fourteen days, to empty the gut. It seems reasonable to assume that this would be distressing. The fish may then be netted or sometimes sucked by pumps through pipes out of the cage or nets in which they have been living, and into a boat or directly to the processing facilities. Salmon pumped for more than two minutes may emerge too exhausted to swim properly, because of their efforts to fight against the turbulence. When nets are used to remove fish from the water, the net can contain hundreds of kilograms of fish, which can cause "bruising, crushing, puncture and abrasion injuries from contact with other fishes," or from contact with the net itself as well as from hard surfaces with which the outside of the net comes into contact.[131] With valuable fish like salmon, the quality of the flesh may be spoiled by rough treatment, so if they are taken to slaughter by boat, it will be a boat with a well that holds water; but carp, tilapia, eels, and loach are routinely transported without any water. They will then undergo what is probably an extreme form of distress from lack of oxygen, as well as from the vibration and the pressure of other fish all around them.

In addition to the 100 billion fish estimated to be intensively farmed each year, many more invertebrate aquatic animals are also farmed—according to one study, 440 *billion* farmed shrimp are killed each year.[132] Although the evidence that shrimp can feel pain is not as strong as that for vertebrates and other crustaceans like lobsters and crabs, this is in part because they have been less studied. The possibility that they are sentient cannot be excluded, and given the vast numbers raised and killed, even a small chance that they can feel pain should not be ignored.[133] A major Spanish corporation is now planning to begin large-scale intensive farming of octopus, highly intelligent animals who are curious and very probably self-aware.[134]

PAINFUL PROCEDURES

FOR MILLENNIA, PAIN HAS BEEN INFLICTED ON ANIMALS FOR HUman benefit whether they are reared by modern or traditional methods and, as the authors of an international study focusing on cows, sheep, and pigs write, "Pain is still neglected, under-recognized, and under-treated in farm animals." They discuss the following routine painful procedures: castration of boars, bulls and rams; cutting away wrinkly folds of skin on sheep that could otherwise lead to flystrike, a procedure known as mulesing; hot-iron or freeze branding, especially of cows; ear tagging and ear notching of calves, lambs, and piglets; tail docking of pigs and sheep; dehorning or disbudding (removing the horn bud so that no horn forms) in calves and goats; and inserting a metal ring through the nose of bulls and sows. All of these procedures can cause severe physical pain, sometimes lasting many hours. Dehorning and disbudding, for example, are described as producing "severe pain for hours as evidenced by severe burns and large open wounds, changes in behavior (e.g., vocalization, kicking, and falling), decreases in mechanical nociceptive thresholds, and increases in serum cortisol levels."[135] Hot-iron branding causes a third-degree burn that takes eight to ten weeks to heal, and tail docking can cause the formation of neuromas that will be painful as long as the animal lives.

Some of these painful procedures could be avoided, for example, by using breeds of cows who do not develop horns, or less wrinkly sheep who do not need to be mulesed. With all the others, the pain could be eliminated or at least greatly reduced by the use of anesthetics or analgesics. Yet all too often, these forms of pain relief are not used. After video of cows being dehorned in Western Australia was shown on television, the state minister of agriculture said that "the community will rightly question how producers can be permitted to operate in this way." On the video, a station worker is heard to say: "Nobody uses painkiller. Too expensive, too much hassle."[136] In the United States, the international study mentioned earlier showed that over 40 percent of veterinarians dehorned cows without using any form of pain relief. Some European countries and New Zealand have rules requiring that pain relief must be used for at least some of these painful procedures.

In the nineteenth century, the cruelty of transportation and slaughter aroused anguished pleas from the humane movement, especially in the United States, where the animals were driven from pastures near the Rockies down to the railheads and jammed into railway cars for several days without food until the trains reached Chicago. In 1906, a federal law was passed limiting the time that animals could spend in a railway car without food or water to twenty-eight hours, or thirty-six hours in special cases. After that time the animals must be unloaded, fed, given water, and rested for at least five hours before the journey is resumed. Obviously, a period of twenty-eight to thirty-six hours in a lurching railway car without food or water is still long enough to cause distress, but it was an improvement.

The 1906 U.S. federal law about transporting animals by rail said nothing about animals being transported by truck. Trucks were not used for transporting animals in those days. When the 1990 edition of this book was published, the transportation of animals by truck was still not regulated at the federal level. Finally, in 1994, the law was amended in a manner that could be interpreted to include trucks. The U.S. Department of Agriculture has subsequently acknowledged that the law does now apply to trucks, but investigations conducted by the Animal Welfare Institute and Animal Outlook indicate that it is routinely flouted. In August 2021, for example, investigators followed a truck packed with pigs for more than thirty hours, in temperatures up to 91°F (33°C), without any stops for the pigs to be given water or to be able to rest or eat. The investigators documented the pigs screaming at night as they were crammed together in the trailer. Despite the acknowledgment from the Department of Agriculture that the twenty-eight-hour law applies to trucks, neither the Animal Welfare Institute nor Animal Outlook is aware of a single prosecution for a violation of the law.[137]

In the European Union, the law does not allow the long uninterrupted journeys that are permitted in the United States. Instead, after fourteen hours in transit, there must be a stop of one hour to rest and provide water for the animals. Then the animals can be transported for a further fourteen hours, but after that, if the journey is not complete, they must be unloaded and rested for twenty-four hours at an approved control post before the journey can be continued. A study of cows coming from France into Italy showed, however, that 21 percent of journeys did not include the compulsory hour stop after fourteen hours, while only 30 percent appear to have complied fully with the regulations.[138]

Animals placed in a truck for the first time in their lives are likely to be frightened, especially if they have been handled hastily and roughly by the men loading the truck. The motion of the truck is also a new experience, and one that may make them ill. After one or two days in the truck without food or water they are desperately thirsty and hungry. If the journey is in winter, subzero winds can result in severe chill; in summer, the heat and sun may add to the dehydration caused by the lack of water. It is difficult for us to imagine what this combination of fear, travel sickness, thirst, near starvation, exhaustion, and possibly severe chill feels like to the animals.

Although the animals cannot describe their experiences, the reactions of their bodies tell us something. All animals lose weight during transportation. Some of this weight loss is due to dehydration and the emptying of the intestinal tract; but more lasting losses are also the rule. One recent study showed that the majority of the weight loss is lost tissue, which suggests severe stress. This "shrinkage," as it is known in the trade, is regarded by researchers as an indication of the stress to which the animal has been subjected. Shrinkage is, of course, a worry to the meat industry, since animals are sold by the pound. The study referred to claims that just a 1 percent reduction in shrinkage among cows sent to feedlots in the United States would provide an economic benefit of more than $325 million for the beef industry.[139]

Animals who die in transit do not die easy deaths. They freeze to death in winter and collapse from thirst and heat exhaustion in summer. They die, lying unattended in stockyards, from injuries sustained in falling off a slippery loading ramp. They suffocate when other animals pile on top of them in overcrowded, badly loaded trucks. They die from thirst or starve when careless stockworkers forget to give them food or water. And they die from the sheer stress of the whole terrifying experience. The animal who you may be having for dinner tonight did not die in any of these ways; but these deaths are and always have been part of the overall process that provides people with their meat.

SLAUGHTER

KILLING AN ANIMAL IS IN ITSELF A TROUBLING ACT. VERY FEW people ever visit a slaughterhouse, and films of slaughterhouse operations are not popular on television. People may hope that the meat they buy came from an animal who died without pain, but they do not really want to know about it. But those who, by their purchases, require animals to be killed do not deserve to be shielded from this or any other aspect of the production of the meat they buy.

There is no need for death to be painful. In countries with humane slaughter laws, death is supposed to come quickly and painlessly. First the animals should be stunned by electric current or a captive-bolt pistol; then, while they are still unconscious, their throats are cut and they bleed to death. In practice, however, many animals have very painful deaths.

Much of the suffering that occurs in slaughterhouses results from the frantic pace at which the killing line must work. Economic competition means that slaughterhouses strive to kill more animals per hour than their competitors. That emphasis on the speed of the killing line is conveyed by the title of Timothy Pachirat's chilling book on industrialized slaughter in America, *Every Twelve Seconds*, which refers to how often a cow was killed at the Omaha slaughterhouse where Pachirat worked incognito for five and a half months in 2004. The pressure to keep the line moving was always there. It led to Pachirat's co-workers telling him that he had to use an electric prod on the cows moving up the chute to the "knocking box," because the line was not tight enough. He describes one of his co-workers using the electric prod constantly, sticking it under the cows' tails and into their anuses, shocking them even when they were so tightly packed that the cow he was shocking had its head between the rear legs of the cow in front of it. The shocks made the cows kick, jump, and bellow.[140]

The imperative to work as fast as possible also meant that when a cow was not properly stunned, what happened next depended on how close to the action a U.S. Department of Agriculture inspector might be. If the inspector was present, the line would be stopped. If no inspector was there, the line would keep going, and the cow might still be conscious when workers began the process of dismemberment, cutting into the tail, leg, and anus. As Pachirat explains, the workers doing this are standing above the suspended cow, and as they cannot see the head, they may not even be aware that they are cutting into a conscious animal. Only those who see the movement of the cow's head and eyes would know this.[141]

Pachirat describes one occasion when a cow fell down in the chute leading to the killing box. Efforts to move the cow by pulling hard on a nose ring failed when the ring ripped through the cow's nostrils. With the panicked cow thrashing around on the floor of the chute, the supervisor instructed the other workers to use their electric prods to drive the cows over the fallen cow, while at the same time telling them to keep a look out for inspectors. The episode ended only when a Department of Agriculture inspector arrived and ordered that the downed cow be stunned and killed, but not processed.[142] Slaughterhouses are not pleasant places to work, and most employees do not last long. When Pachirat phoned slaughterhouses to enquire about getting work there, the general response was along the lines of "Just show up in person. We're always hiring." This response, he says, was consistent with the statistics he had read, with some estimates putting the average annual turnover rate at more than 100 percent, meaning that the average employee doesn't last a year in the job.[143] This results in a constant stream of inexperienced staff handling frightened animals, and always under constant pressure to do it as fast as possible.

The incidents that Pachirat describes happen at slaughterhouses in every country, no matter how strict the laws may be. Animal Equality UK has obtained videos of conscious animals undergoing extreme suffering in slaughterhouses for pigs in Italy, for cows in Brazil, for salmon in Scotland, for sheep in Wales, and for turkeys in Spain.[144] In 2019, an investigation by the French animal advocacy organization L214 of the Sobeval slaughterhouse for veal calves, in Dordogne, produced video of incidents so sickening that L214 had to produce a toned-down version that is less difficult to watch.[145] If you eat meat, you may hope that the animals whose flesh you are eating died quickly and humanely, but you cannot really know how they died.

Bad as the killing of these land animals is, the slaughter of fish is even worse. There is no humane slaughter requirement for wild fish caught and killed at sea, nor, in most places, for farmed fish either. Fish caught in nets by trawlers are dumped on board the ship and, if not crushed by the sheer weight of the fish on top of them, are allowed to suffocate. Large fish like tuna may be killed by gaffing: that is, hoisting them out of the water with a hooked or barbed spear, then thrusting a spike into the brain of the fully conscious animal. Impaling a live bait fish on a hook is a common commercial practice, used in long-line fishing with hundreds or even thousands of hooks on a single line that may be 50 to 100 km (30 to 60 miles) long. When fish take the bait, they are likely to remain caught for many hours before the line is hauled in. Gill nets are walls of fine netting in which fish become snared, often by the gills. They may suffocate in the net, because with their gills constricted, they cannot breathe. If not, they may stay there for many hours before the nets are pulled in. Deep sea fish hauled to the surface are likely to die from decompression, resulting in internal organs like their swim bladders bursting. And the scale of all this is vast, with a range of 787 billion to 2.3 trillion vertebrate fish killed every year.[146]

For farmed fish, the least painful end that any fish is likely to receive at human hands is to be stunned, either by a blow to the head or electrically, before being cut open and disemboweled. Few are so lucky. Some will be placed on ice, which may lead to loss of consciousness, although not instantly and not in all cases. They will then die by asphyxia—in other words, suffocating because they are in air and cannot breathe—or by having their gill arches slit open and bleeding to death. Catfish have their heads cut off. In the Netherlands, the majority of farmed fish are electrically stunned before being killed, but in most other countries they are not.[147]

If the slaughter of fish is, as I've suggested, typically worse than the slaughter of land animals, the use of salt to kill freshwater fish may be even worse than most other ways of killing fish. Fish without a covering of scales, such as eels and loaches, are especially sensitive to salt because the concentrated salt solution sucks moisture out through the fish's skin, a process known as osmotic dehydration. The European Food Safety Authority asked its Panel on Animal Health and Welfare to provide a scientific opinion on this method, which is commonly used in Europe to kill eels. The panel reported:

> Eels make extremely vigorous attempts to escape from salt . . . and take a long time to lose consciousness. . . . Based on behavioral data and eels' reaction to stimulation, it may even take more than 25 min.[148]

The pain suffered by fish undergoing death by osmotic dehydration may be akin to the pain we feel if we get salt in our eye, because that also dehydrates our eye.

The use of salt to kill fish has been banned in Germany, the Netherlands, and New Zealand, with some other countries likely to follow their lead.[149] But in East Asia, this method of killing is used on loaches, small freshwater fish of which at least 18 billion are reared and killed every year, more than any other fish species and about one fifth of all fish produced by aquaculture worldwide. The vast majority of loaches are produced and consumed in China, with significant numbers also eaten in Japan and South Korea. In these countries fish have no legal protection against cruelty, and there also appear to be few cultural inhibitions against causing fish to die slowly and painfully. Hence animal advocates do not need to go undercover to make videos showing how much loaches suffer. They just have to go to YouTube, where videos showing very clearly the agony of the loaches are posted and commented upon by people interested in food, not animal welfare. A video made by JJin Food, for example, promotes Korea's "famous" *chueotang*, or loach soup. It shows in detail the delivery to a kitchen of thousands of live fish in plastic buckets. The loaches are then poured into a huge bowl and sprinkled with salt, which causes them to writhe violently. The camera cuts away and returns, we don't know how much later, to show the fish, still writhing but now covered with a white slime, being transferred to a large mixing bowl that stirs them around and washes the slime off them. Then, with some of them still clearly alive and moving, they are tipped into a cauldron of steaming water.[150] In another video that shows the same process, a Korean worker says: "It's painful when you have salt in it, but to make a delicious *chueotang*, you have to give it some pain."[151] You could, though, avoid causing the pain by instead eating a delicious vegan version of *sundubu jjigae* (a Korean tofu soup, easily prepared from entirely plant-based ingredients).

CHAPTER 4

Living Without Speciesism . . .

while combating climate change and enjoying a healthier life

EFFECTIVE ALTRUISM FOR ANIMALS

NOW THAT WE HAVE UNDERSTOOD THE NATURE OF SPECIESISM and seen the consequences it has for nonhuman animals, you will, I hope, be asking yourself: What can I do about it?

In the first decade of this century, Toby Ord and Will MacAskill, then graduate students in philosophy at Oxford University, began thinking, researching, and writing about how to do the most good they could. Their work sparked a movement that became known as effective altruism.[1] Effective altruists make it one of their goals in life, among others, to make the world a better place, and they use reason and evidence to find out how to do that as effectively as possible. Most effective altruists recognize that reducing animal suffering is one important way of doing good. By thinking like an effective altruist, you may be able to do more good for animals, and for humans too, than you otherwise would.

There are many things you can and should do about speciesism. Some of the obvious things are to make your friends aware of these issues, educate your children to be concerned about the welfare of all sentient beings, and protest publicly on behalf of nonhuman animals whenever you have an effective opportunity to do so. Depending on your financial situation, you may be able to donate to charities campaigning to reduce animal suffering. There are now nonprofit organizations that provide independent assessments, available online, of which charities are the most effective in reducing animal suffering. (Animal Charity Evaluators is the best-known.) Distressingly, from an effective altruism perspective, the animal charities that receive the most money from the public are devoted to helping dogs and cats, whereas—as you already know from reading the previous chapters—the suffering caused to animals by factory farming is far greater and your money will do much more good if donated to organizations trying to improve the lives of animals confined in factory farms, or those seeking to reduce the number of animals in factory farms by encouraging people not to buy their products.

Effective altruism also applies to career choice. If you have not yet chosen your career or are in a position to switch to a different career, think about one that could have an impact on reducing animal suffering. There are many possibilities. You could work for one of the charities rated highly by Animal Charity Evaluators. People with the ability and the commitment to be excellent campaigners or managers are in short supply. So too are senior researchers. But if you would find that kind of work too emotionally draining, you can also help by pursuing a career that enables you to earn enough to help fund the campaigns of those charities. Should you earn well and live modestly, your donations may enable the organizations to expand and hire additional people. Whatever they achieve for animals would not have happened without your financial support. A high-stakes/high-risk option is to go into politics and make the protection of animals one of your policy goals. Very few of those who enter politics to change the world succeed in reaching a position from which they can make a major positive difference; but the few who do get there may have the extraordinary opportunity of improving the lives of billions of animals and people.[2]

EATING ETHICALLY

ALL OF THE ACTIONS JUST MENTIONED ARE IMPORTANT THINGS to do, but there is one more step we can take that underpins, makes consistent, and gives meaning to all our other activities on behalf of animals: We can take responsibility for our own lives, and make them as free of cruelty as we can. We can, as far as is reasonable and practical in our individual circumstances, stop buying and consuming meat and other animal products.

Switching to a plant-based diet is not merely a symbolic gesture. Nor need it be an attempt to isolate oneself from the ugly realities of the world, to keep oneself pure and so avoid responsibility for the cruelty and carnage all around. Avoiding animal products is a practical and effective step toward a healthier lifestyle for yourself that will also have other, even more important consequences. It can contribute to reducing the misery we inflict upon animals, to making *more* food—not less—available to feed a growing world population, to saving the planet from catastrophic climate change, and to lessening the risk of another pandemic.

The people who profit by exploiting large numbers of animals do not need our approval. They need our money. The purchase of the flesh, milk, and eggs of the animals they rear is the most important form of support that factory farmers seek from the public (although in many countries, government subsidies certainly help). Giant agribusiness corporations will use intensive methods of animal production as long as they can sell what these methods produce. Their profits give them the resources needed for blocking far-reaching change and they will continue to defend themselves against criticism by saying that they are only meeting the demand from the public.

Hence the need for each one of us to stop buying the products of the animal factories. Until we do that we are contributing to their continued existence, prosperity, and growth, and all the cruel practices used in rearing animals for food. In the supermarket and at the dinner table, we have an opportunity to do something more than merely talk about how someone else should do something to stop the cruelty. It is easy to take a stand against bullfighting in Spain, eating dogs in South Korea, kangaroo shoots in Australia, the Canadian seal slaughter, or the bloody massacre of dolphins in Japan; but we reveal our real values when the issue comes into our own home. If we are continuing to eat chicken produced by breeding birds who are in pain because their legs cannot support their fast-growing bodies, or meat from pigs kept confined indoors all their lives, it is hypocritical of us to object to eating dogs. It is when we eat that the consequences of speciesism intrude most directly into our lives. Boycotting the factory production of animals enables us to attest to the depth and sincerity of our concern for nonhuman animals.

To make this boycott as effective as possible, we must not be shy about what we are refusing to eat and why. I first learned of the existence of factory farming from Richard Keshen, a fellow-student at Oxford University with whom, one fateful day, I had lunch. He asked whether a dish we were offered had meat in it and, when told that it did, chose something else. I asked him what his problem with meat was. Within a month of that initial conversation, I became a vegetarian. Being asked about your food choices is an opportunity to tell people things they may not know about how the animals they are eating were treated. One conversation changed my life, and my views have since influenced many others. If a boycott can reduce animal suffering, then we must encourage as many as possible to join in, and we can't do this unless we ourselves set the example.

If avoiding factory farm products is a form of boycott, then what do we do if the boycott isn't working? That question has to be asked, because since I called on readers to boycott meat in the first edition of this book, worldwide consumption of meat has increased from 112 million tons to more than 300 million tons, with virtually all of the additional meat coming from factory farms.[3] A large part of that increase is due to the world's population having doubled in size during that period, and most of the rest is the result of an otherwise welcome reduction in poverty, especially in Asia. Meat is expensive, and so people consume it only when they can afford to do so. China's per capita meat consumption tripled between 1990 and 2021, and Vietnam's quadrupled over the same period, while there were also sharp increases in Brazil, India, Indonesia, Mexico, Pakistan, and South Africa. Countries that were already affluent in 1990 did not have such a clear trend, with moderate increases in Australia, Israel, Norway, and Japan, more modest increases in the United Kingdom and United States, and decreases in Canada, New Zealand, and Switzerland. In a promising sign, meat consumption in both Germany and Sweden declined by more than 12 percent in the period from 2011 to 2019.[4]

Clearly, my call for a boycott of meat has been a dismal failure. Does that mean that, for people who are persuaded by the case I make in this book, there is no point in continuing it?

With indirect boycotts, that may be the case. For example, when People for the Ethical Treatment of Animals asked its supporters to boycott Air France because it was transporting thousands of wild-caught monkeys to laboratories, that boycott was either going to succeed in getting Air France to change its policies regarding transporting primates to laboratories, or it was going to fail and achieve nothing. (Fortunately, it succeeded.)[5] Boycotts aiming at changing the conduct of a company or a government are often like that, win or lose, with little ground in between those extremes. If you are boycotting one product—in this example, passenger flights—that a company sells in order to persuade that company to stop doing something completely different, like transporting monkeys, it would be reasonable to say that if you aren't succeeding, there is no point in continuing the boycott.

Boycotting factory farms is different because it is a direct boycott of the product to which the boycotters object, and hence it has a direct impact on the amount of suffering that the boycotters are trying to prevent. The number of animals that factory farms raise and slaughter increases or falls directly in response to the demand for those products. To cease to support the broad movement against buying them would only lead to more animals leading miserable lives on factory farms. Even if the trend is not going in the direction we hope, that's no reason to make things even worse.

How much difference can an individual make? One way of attempting to answer this question is to try to calculate the total number of animals killed for food and divide it by the total meat-eating population. Maria Salazar, of Animal Charity Evaluators, did this, restricting her calculations to vertebrates only. She concluded that each year, the average consumer of a plant-based diet saves approximately 79 wild-caught fishes, 14 farmed fishes,

11.5 farmed birds, and half a farmed mammal, for a total of 105 vertebrates.[6]

Despite such calculations, some have questioned whether our individual purchasing decisions really can make a difference to the number of animals raised and killed. They point out that one person's decision to avoid buying, say, a factory-farmed chicken corpse at a supermarket isn't likely to make any difference to the number of chickens raised and killed. That's because supermarket purchasing policies are not so fine-tuned that one fewer chicken sold this week will mean a reduced number of chickens ordered next week, let alone a reduced number of chicks hatched and transported to the growers who raise them.[7] But the laws of supply and demand dictate that a lower demand for chicken corpses must, eventually, lead to a lower order from the supermarket and thus a lower number of chickens raised. There must be a threshold at which the supermarket's order will rise or fall, and to ignore the chance that your purchase will trigger this change is to ignore the "expected value"—or rather, in this case, "expected disvalue"—of your purchase. To illustrate: Suppose that every day you can choose between receiving $1 or receiving a lottery ticket that gives you a 1 in 100 chance of winning $100. If you want to maximize your wealth over the long run, it is just as good to opt for the lottery ticket because it has an expected value of $1 (the prize divided by the odds against getting it), which is the same as the certainty of gaining $1. Similarly, if your goal is to reduce the suffering of chickens, then you should think of the expected value of not buying dead chickens. Suppose the supermarket orders chickens only in hundreds, changing its weekly order when the number sold reaches or falls below a number that is a round hundred. Now suppose that in Week 1, I decide not to buy a chicken, and that means that this week the supermarket sells 8,721 chickens rather than 8,722. Then my purchase will not affect the supermarket's order. This may continue for many more weeks, until one week, because I do not buy a chicken, the supermarket sells 8,699 chickens rather than 8,700. At the end of this week, therefore, the supermarket orders 100 fewer chickens. Now I have been personally responsible for not just one fewer chicken being ordered but for one hundred fewer. A 1 in 100 chance of saving 100 chickens from pain and suffering has the same expected utility as the certainty of saving one chicken, so the fact that the supermarket's purchasing policies are "lumpy" in this way doesn't make any difference to the expected utility of my decision to not buy chicken. This point will also apply to the decisions of the chicken supplier to hatch more chickens, which in turn will depend on the supermarket orders. It doesn't matter whether the threshold for change is a hundred, a thousand, or a million. The expected utility is just the same as it would be if there were a direct one-to-one relationship between the number of chickens bought and the number raised. Hence we can assume that our purchasing choices, together with those of the many others who are avoiding factory-farmed animal products, are reducing the number of animals raised in factory farms.

Two economists, Bailey Norwood and Jayson Lusk, provide some empirical confirmation of this idea in *Compassion, by the Pound: The Economics of Farm Animal Welfare*. They allow for a further complication: A decline in demand for a product may lead to lower prices, and that in turn will mean that others buy more. Nevertheless, there is a limit to how much anyone can increase their food consumption. According to their calculations, if there is a drop in the demand for eggs equivalent to the eggs produced by 100 hens, 91 fewer hens will be confined in egg factories; if 100 fewer chickens are bought, 76 fewer will be raised in chicken factories; for 100 fewer pigs bought, 74 fewer pigs will be living in pig factories; for 100 cows raised for beef bought, 68 fewer are going to be raised and slaughtered; and if the milk of 100 cows is not bought, 56 fewer cows will be kept for their milk.[8]

FEEDING THE WORLD

AGRIBUSINESS LEADERS SOMETIMES SAY THAT WE NEED TO HAVE factory farms to feed the world's growing population. That is the opposite of the truth. Because the truth here is so important—important enough, in fact, to amount to a convincing case for plant-based eating that is quite independent of the concern for the welfare of the animals—I need to say something about this misconception.

Every animal raised by humans for food has to eat in order to grow to the size and weight at which it is killed and eaten. If a cow gains weight grazing on grass growing on land that cannot produce a crop we can eat, the result will be a net gain of available food for humans, since the cow provides us with protein that we cannot readily extract from grass. But if we take cows and place them in a feedlot, we need to bring food to the cows, and that food must come from crops grown on arable land that could be used to grow crops we can eat. Cows, no matter how strictly we prevent them from exercising, must still burn food merely to keep their bodies warm and maintain the vital functions that keep them alive and to build bones and other body parts we do not eat. Only a small part of the nutrients we feed cows remains to be turned into flesh that, if not wasted, is eaten by humans.

Most people have no idea how high a proportion of the crops we grow are fed to animals. When I tell people about my favorite ways of cooking tofu, they often say, "But tofu is made from soy, and the Amazon is being cleared to grow soybeans, so I wouldn't want to support that!" What they don't realize is that 77 percent of global soy production is fed to animals and thus converted to meat and dairy products. Some is used for biofuel, and some for vegetable oils. Only 7 percent is eaten directly by humans, as tofu, tempeh, edamame, and soy milk. It's not tofu that is driving deforestation but the meat and dairy industries.[9]

Just how wasteful is this process? Because animal products provide us with both calories and protein, we need to divide that question into two: How many calories do we feed to animals in order to get each calorie of food from the animal product? And how many grams of protein do we feed to animals in order to get one gram of protein from the animal product? One study, looking at the United States only, provides the following answers:

- Beef, the least efficient product, has a caloric conversion efficiency of only 2.9 percent and a protein conversion efficiency of 2.5 percent. In other words, we have to feed cows 34 calories in order to get 1 calorie we can consume, while with protein the ratio is 40:1.
- Pork is more efficient, at 9 percent for both caloric and protein conversion efficiency, but that still means we are feeding them more than 10 units for each one we obtain from them.
- Chicken is another step up, at 13 percent caloric and 21 percent protein efficiency.
- Cows producing milk rather than meat have a caloric conversion efficiency of 17 percent and a protein conversion efficiency of 14 percent.

- Hens kept for their eggs have a caloric conversion efficiency of 17 percent and, at 31 percent, the best protein conversion efficiency of all of these animal products.

The most efficient conversion of any form of animal food eaten on a large scale by Americans, therefore, still involves feeding more than 3 grams of protein to hens for every gram of protein we get from the eggs they lay. The authors of this study pointed out that a dietary shift in the United States that replaces just beef with plant-based high-protein foods such as legumes would have the potential to provide enough food for an additional 190 million people. In a separate study, the same authors estimated that replacing all animal-based items in the U.S. diet with nutritionally equivalent plant-based alternatives would free enough land to feed an additional 350 million people.[10]

Global patterns of food consumption vary greatly in the amount of land they require to feed people. Farmed animals occupy nearly 80 percent of the world's agricultural land, while producing less than 20 percent of the global supply of calories. Researchers from Scotland, Germany, and Australia found that if everyone in the world ate like people in India, 55 percent less land would be required to feed them; whereas if everyone ate like people in the United States, we would need 178 percent more land. The difference, of course, depends on the types of food eaten—and particularly on the impact of animal products—rather than on the quantities eaten. There is, quite rightly, concern about spoilage, waste, and other losses in the passage of food from farm to table, but by far the greatest form of waste is using arable land to grow food for animals. Waste and overconsumption in the United States play a much smaller role than the choice to eat more animal products instead of consuming plants directly.[11]

Ocean fishing does not use land, of course, but it still distorts the global distribution of food in favor of the rich. According to the UN Food and Agriculture Organization, marine fishery resources are declining, with the fraction of biologically sustainable fishery stocks decreasing from 90 percent in 1974 to 65 percent in 2019.[12] When fish of a particular species and locality are being caught faster than they can reproduce, it becomes impossible to sustain that level of fishing and so the catch is bound to fall; if fishing continues unabated, the fish will become "commercially extinct"—that is, there will no longer be sufficient fish to make commercial fishing in that area viable. This has happened to several once-abundant species of fish.

Modern fishing fleets trawl the fishing grounds systematically with fine-gauge nets that catch everything in their path. Because trawling involves dragging a huge net along the previously undisturbed bottom of the ocean, it damages the fragile ecology of the seabed. In addition to the disruption of ocean ecology caused by all this overfishing, there are tragic consequences for humans too. In many poorer countries, small coastal villages that for millennia lived by fishing are finding that the fish stocks that were their traditional source of protein and income have vanished. This has caused a crisis in West Africa, where illegal and unregulated fishing has caused the collapse of coastal fishing industries. It has also contributed to the number of Africans desperately trying to get to Europe—which, ironically, is where most of the fish caught by the trawlers that destroyed their local fishing grounds are sold.[13] The fishing industry of the developed nations has become one more form of redistribution from the poor to the rich. This redistribution is especially glaring when the fish caught off the coasts of poorer nations are fed to farmed carnivorous fish like salmon to create a product that only the affluent can afford.

CONSCIENTIOUS OMNIVORES

I HOPE THAT ANYONE WHO HAS READ THIS FAR WILL RECOGNIZE the moral necessity of refusing to buy or eat the flesh or other products of birds and mammals who have been reared in factory farm conditions. To see that supporting factory farming is wrong doesn't even require the rejection of speciesism nor acceptance of the principle of equal consideration of animals. Joel Salatin, for example, a guru of sustainable animal raising whose Polyface Farm, in Virginia, was praised by Michael Pollan in his bestselling book *The Omnivore's Dilemma*, denies all notions of equality between humans and animals but condemns the "egregious, deplorable" conditions of factory farming in which a chicken is "living in its toilet all the time."[14] Avoiding factory farming products is the clearest case of all, the absolute minimum that anyone with the capacity to look beyond considerations of narrow self-interest should be able to accept.

Let us see what this minimum involves. A study by the Sentience Institute estimated that in the United States, over 99.9 percent of chickens raised for meat are kept in factory farms, 99.8 percent of turkeys, 98.3 percent of pigs, 98.2 percent of egg-laying hens, and 70.4 percent of cows. The study also says that "virtually all U.S. fish farms are suitably described as factory farms."[15] To be a truly conscientious omnivore, therefore, requires some investigation into where one's animal products are coming from. It means that, unless we can be sure of the origin of the particular item we are buying, we should avoid chicken, turkey, meat from pigs, eggs, veal, and beef. At the present time fewer than 1 percent of sheep are kept intensively, so lamb and mutton are unlikely to be from factory farms.[16] The likelihood of beef coming from a feedlot will depend on where you live. In the United States, only

1 percent of beef comes from cows that stay on grass their entire lives, whereas in Australia, half of all beef comes from cows kept on pasture.[17] It is possible to obtain supplies of all these meats that do not come from factory farms, though often at much higher prices. In the case of chickens raised to be eaten, the price difference is so great that very few farmers now raise them outdoors. Bear in mind too that "cage-free" eggs are likely to come from hens kept indoors in crowded conditions. In the United States, even the label "organic" does not guarantee that hens have access to the outdoors. If you are going to buy eggs, look for the words "free-range" or "pasture-raised" eggs. In the European Union and the UK, egg cartons must be stamped with a code: 0 for organic (which in those jurisdictions does require access to free-range), 1 for free-range, 2 for cage-free but indoors, and 3 for eggs from caged hens. In Australia, eggs must be labeled free-range, barn-laid, or from caged hens. When buying free-range or pasture-raised eggs, it is worth checking how much space the birds have, because if there are too many birds on an outside run, the grass will turn into packed dirt. Some free-range producers state their stocking density on the egg carton or on their website. If they don't, contact them and ask. Stocking densities of 600 birds per acre (or 1,500 per hectare) will enable the birds to spread out and permit grass to grow in favorable weather.

If you agree with the argument thus far and stop eating factory-produced chicken, pigs, calves, and cows, as well as the milk and eggs from factory-kept cows and hens, the next question is whether to go vegetarian or to become a conscientious omnivore. Conscientious omnivorism is defended by, among others, Joel Salatin, Michael Pollan, and Roger Scruton, a fox-hunting English philosopher who died in 2020. In 2005, during a debate Scruton and I had at Princeton University, he told the audience that he had a farm on which he raised pigs and that he had named one of them Singer. Soon, he remarked, he would kill and eat Singer who, he emphasized, would not have existed and had the life that he was at that very moment enjoying if no one ate meat. A similar point had been made more than a century earlier by Leslie Stephen (the father of the novelist Virginia Woolf), who wrote, "The pig has a stronger interest than anyone in the demand for bacon. If all the world were Jewish, there would be no pigs at all."[18]

Conscientious omnivores oppose factory farming but continue to eat animal products from farmers who treat their animals well. Scruton argued that painlessly killing animals is not in itself wrong, because an animal's death cannot be "untimely" in the way that a human death often is. For humans, to have one's life cut short can be a tragic waste, because the person who has died had more to achieve. In contrast, Scruton claimed, animals do not have achievements, and so "to be killed at thirty months is not intrinsically more tragic than to be killed at forty, fifty, or sixty." Killing and eating domestic animals is justified, he claimed, when they "are properly looked after, when all duties of care are fulfilled, and when the demands of sympathy and piety are respected."[19]

In the first edition of this book, I rejected Leslie Stephen's argument on the grounds that it requires us to think that bringing a being into existence confers a benefit on that being—and to hold this, we must believe that it is possible to benefit a nonexistent being. This, I wrote, was nonsense; but now I am less sure that it is. After all, most of us would agree that conceiving a child who we know will have a genetic defect that would make their life painful and short would harm the child. Yet if we can harm a nonexistent child, surely we can also benefit a nonexistent child. To deny this, we would need to explain the asymmetry between the two cases, and that is not easy to do.[20]

Views about whether painlessly killing animals is intrinsically wrong are relevant here. I have not based my argument on the claim that it is wrong to kill animals, because probing that claim leads to much more difficult philosophical questions than the wrongness of causing or supporting avoidable suffering. Scruton, as we have seen, thought that killing animals is not tragic because even if they were to live longer, they would not achieve anything. One might question the factual claim: Isn't giving birth to, and successfully feeding and rearing, a calf an achievement for a cow? Scruton might have responded by saying that to count as an achievement, one must consciously plan and work for a goal, which cows, arguably, do not do. We could resist such attempts to draw a line between humans and other animals, but let us grant, to see where the argument then goes, that cows do not have achievements in the same way that most humans do.

We could still reject the moral claim Scruton makes. We might think that it is wrong to kill a being who is capable of enjoying their life, whether or not this prevents them from achieving anything. In support of this broader view of when killing is wrong, we could argue that the distinction between beings who have achievements and those who do not would permit the killing of humans with cognitive capacities no greater than those of the pigs Scruton killed and ate—and perhaps even of couch potatoes who spend their lives passively absorbing whatever is on television. They too may have no prospect of future achievements.

It is true that the humans I have just mentioned do not owe their existence to us, in the way that the domestic animals we breed to raise, kill, and then eat do. But what if they did? Kazuo Ishiguro, in his 2005 novel *Never Let Me Go*, imagines a world in which people clone themselves so that the clones can later serve as organ donors for their aging bodies. The clones live good lives, into their teenage years, and have a painless death, usually without learning what is going to happen to them. (In the novel, some of them do learn what their fate will be, but to keep the parallel with animals we can focus on those who do not.) If killing humans for their organs were prohibited, the cloned people would not have existed at all. Does this mean that killing them for their organs is justifiable? If we find this idea abhorrent, then can we justify the parallel view about animals who would not have existed if we were not permitted to kill them, and whose lives are, on the whole, good?

Henry Sidgwick, whom I regard as the most careful thinker of the nineteenth-century utilitarians, discussed the question of whether it is good to bring more people into existence if they can be expected to live happy lives and will not reduce the happiness of others so much that the total amount of happiness in the world will be lower. He thought that under these conditions, bringing more people into existence would be a good thing to do. He may therefore have supported the views of Scruton and other conscientious omnivores, for if it is good to bring happy humans into the world, other things being equal, the same should hold for bringing happy animals into existence.

When I was a graduate student at Oxford, I attended a celebrated series of seminars in which Derek Parfit, one of the most highly regarded philosophers of the past fifty years, argued that Sidgwick's view that we should do what will bring about the greatest total amount of happiness leads to a "repugnant conclusion." The idea Parfit considered repugnant is that a world with a huge population of people whose lives are just barely worth living could have a greater total amount of happiness—and thus be better—than a world with a much smaller population of people leading lives of the highest possible quality. Before you reject Sidgwick's total view because of this implication, though, you need to know that Parfit also showed that alternative ethical approaches to decisions affecting who will come into existence have implications that are no less counterintuitive than the repugnant conclusion. Other philosophers have also failed to come up with a satisfactory answer to the problem.[21]

Given the difficulties that I and many other philosophers have with these issues, I remain in doubt whether it is good to bring into existence beings who can be expected to live happy lives and whether this can justify killing them. Somewhat to my chagrin, I admit, I am unable to provide any decisive refutation of the conscientious omnivore. But in practice, even if we admit that we can benefit animals by bringing them into existence, and accept that killing them at an early age is not wrong, meat from animals treated sufficiently well is going to be scarce. Scruton, who raised his own pigs, and—let's assume—killed them painlessly on the farm, may have been able to be confident that they had good lives. Commercial producers who have their animals rounded up and taken from the environment they have known all their lives, to be packed tightly into a truck with other animals, transported to a slaughterhouse, and killed there, are putting them through an ordeal that may outweigh months of keeping them in acceptable conditions.

How should these facts affect our views about conscientious omnivorism? Adam Lerner, a contemporary American philosopher, points out that, as I have already mentioned, we regard it as wrong to knowingly and avoidably bring into existence a child who will have a miserable existence. Nor would we think, Lerner claims, that one could right this wrong by also having a child who will have a good life. Similarly, he argues, people who bring animals into existence to raise them commercially cannot justify that activity on the grounds that some of them will have positive lives, because many farm animals have negative lives, even on farms that allow them to go outside. That's because mothers and their young will often be separated soon after birth, castration may be performed without anesthetic, animals will be branded or have their ears notched, and social groups are likely to be broken up. Finally, except in the rare cases of on-farm slaughter, all the animals will have to endure the ordeal of transportation to the slaughterhouse. Hence Lerner's view is that we should not be conscientious omnivores.[22]

Oliver Goldsmith, the eighteenth-century humanitarian essayist, wrote of people who are opposed to cruelty to animals but are not vegetarian: "They pity, and they eat the objects of their compassion."[23] As this suggests, it is difficult to be consistent in one's concern for nonhuman animals while continuing to dine on them. Our eating habits are dear to us and not easily altered. There is a danger that despite our compassion, we will come to regard pigs, cattle, chickens, and fish as things for us to use. If we find that it is necessary to squeeze these animals' living conditions a little in order to continue to obtain their flesh at prices we are willing to pay, we will be more readily persuaded that these changes do not really harm the animals. No one in the habit of eating an animal can be completely without bias in judging whether the conditions in which that animal is reared cause unnecessary suffering.

In free-market conditions, it will always be difficult to rear animals for food on a large scale without inflicting considerable suffering. The factory farm is nothing more than the application of technology and market forces to the idea that animals are means to our ends.

Moreover, even if farming that gives most of the farmed animals good lives, up to the time when they are killed, can be done on a small scale, it could not feed today's urban populations with the quantities of meat and other animal products they are in the habit of consuming. Accepting and acting on the simple general principle that we should avoid killing animals for food except when it is necessary for our health or survival may have a better outcome in the long run, even if it means that fewer animals come into existence to have short but good lives. At the same time, those of us who choose to be vegetarian or vegan can regard conscientious omnivores as allies in the struggle for a world in which all animals are respected and allowed to live in accordance with their nature and their social needs.

As we are about to see, omnivores who are truly conscientious need to consider not only animal suffering but also the harm that their consumption of animal products does to other humans and to the ecology of the entire planet, because of the greenhouse gases emitted. That is sufficient reason for them to eat more plants and avoid meat, especially from ruminants such as cows and sheep, and dairy products.

CLIMATE CHANGE

I WASN'T AWARE OF CLIMATE CHANGE UNTIL THE 1980S—HARDLY anyone was—and even when we recognized the dire threat that burning fossil fuels posed, it took time for the role of animal production in warming the planet to be understood. Today, though, the fact that eating plants will reduce your greenhouse gas emissions is one of the most important and influential reasons for cutting down on animal products and, for those willing to go all the way, becoming vegan. Here I will, very briefly, report what the best available science indicates about the contributions of animal agriculture to climate change.

The most authoritative reports on climate change are the assessment reports issued by the Intergovernmental Panel on Climate Change, or IPCC. The IPCC was established in 1988 by the

United Nations Environmental Program and the World Meteorological Office with the task of reviewing the state of knowledge about the science of climate change, as well as its economic and social impact and possible strategies for responding to it. In 2022, the IPCC began releasing, in sections, its Sixth Assessment Report. The scale of the enterprise can be judged by the fact that the section on "Mitigation of Climate Change" alone runs to nearly three thousand pages. IPCC reports have many authors, all of them expert in a relevant field, and are based on scientific literature that has been published or accepted for publication, largely though not exclusively in peer-reviewed journals. The draft first undergoes expert review, followed by government reviews. The result is an unusually authoritative document, every sentence of which has had to pass scrutiny from experts searching for errors. If it does err, it is on the side of being too conservative. It is therefore significant that the report on the mitigation of climate change says: "Diets high in plant protein and low in meat and dairy are associated with lower GHG emissions." In the report this sentence is in bold, which is used to highlight statements of particular importance, and the statement is marked as one that is made with "high confidence," another feature of the reports. Later sections of the report discuss strategies for mitigating climate change, and reducing the consumption of meat, especially red meat, is frequently mentioned in those discussions.[24]

In documents that are less constrained than the IPCC reports, the imperative for a change of diet comes through more strongly. "Our appetite for meat is a major driver of climate change," states a summary of *Changing Climate, Changing Diets*, a report from Chatham House, also known as the Royal Institute of International Affairs, in London, which notes that animal agriculture is already responsible for just under 15 percent of global greenhouse gas emissions, about the same as tailpipe emissions from all the world's vehicles. Whereas clean energy has the potential to reduce the use of fossil fuels, however, global meat consumption is forecast to grow by 76 percent by midcentury. That means it will become an increasingly large part of greenhouse gas emissions, leading the authors of the report to conclude that without a reduction in global meat consumption, we will not be able to keep global warming below the "danger level" of 2°C that negotiations such as the Paris Agreement have set as a level that must not be exceeded.[25] Even that level will not be enough to save low-lying Pacific island countries from rising sea levels that will force them to become climate refugees. Experts are arguing, and with good reason, for keeping warming below 1.5°C.

A few years ago, eating locally—eating only food produced within a defined radius of your home—became the thing for environmentally conscious people to do, to such an extent that "locavore" became the *Oxford English Dictionary*'s "word of the year" for 2007. If you enjoy getting to know and support your local farmers, of course, eating locally makes sense. But if your aim is, as many local eaters said, to reduce greenhouse gas emissions, you would do much better by thinking about what you are eating rather than where it comes from. That's because transport makes up only a tiny share of the greenhouse gas emissions from the production and distribution of food. With beef, for example, transport is only 0.5 percent of total emissions. So if you eat local beef you will still be responsible for 99.5 percent of the greenhouse gas emissions your food would have caused if you had eaten beef transported a long distance. On the other hand, if you choose peas you will be responsible for only about 2 percent of the greenhouse gas emissions from producing a similar quantity of local beef. And although beef is the worst food for emitting greenhouse gases, a broader study of the carbon footprints of food across the

European Union showed that meat, dairy, and eggs accounted for 83 percent of emissions, and transport for only 6 percent.[26]

More generally, plant foods typically have far lower greenhouse gas emissions than any animal foods, whether we are comparing equivalent quantities of calories or of protein. Beef, for example, emits 192 times as much carbon dioxide equivalent per gram of protein as nuts, and while these are at the extremes of the protein foods, eggs, the animal food with the lowest emissions per gram of protein, still has, per gram of protein, more than twice the emissions of tofu. Animal foods do even more poorly when compared with plant foods in terms of calories produced. Beef emits 520 times as much per calorie as nuts, and eggs, again the best performing animal product, emit 5 times as much per calorie as potatoes.[27]

Favorable as these figures are to plant foods, they leave out something that tilts the balance even more strongly against animal foods in the effort to avoid catastrophic climate change: the "carbon opportunity cost" of the vast area of land used for grazing animals and the smaller, but still very large, area used to grow crops that are then fed—wastefully, as we have seen—to confined animals. Because we use this land for animals we eat, it cannot be used to restore native ecosystems, including forests, which would safely remove huge amounts of carbon from the atmosphere. One study has found that a shift to plant-based eating would free up so much land for this purpose that seizing the opportunity would give us a 66 percent probability of achieving something that most observers believe we have missed our chance of achieving: limiting warming to 1.5°C. Another study has suggested that a rapid phaseout of animal agriculture would enable us to stabilize greenhouse gases for the next thirty years and offset more than two thirds of all carbon dioxide emissions this century. According to the authors of this study: "The magnitude and rapidity of these potential effects should place the reduction or elimination of animal agriculture at the forefront of strategies for averting disastrous climate change."[28]

Climate change is undoubtedly the biggest environmental issue facing us today, but it is not the only one. If we look at environmental issues more broadly, we find further reasons for preferring a plant-based diet. The clearing and burning of the Amazon rainforest means not only the release of carbon from the trees and other vegetation into the atmosphere, but also the likely extinction of many plant and animal species that are still unrecorded. This destruction is driven largely by the prodigious appetite of the affluent nations for meat, which makes it more profitable to clear the forest than to preserve it for the indigenous people living there, establish an eco-tourism industry, protect the area's biodiversity, or keep the carbon locked up in the forest. We are, quite literally, gambling with the future of our planet for the sake of hamburgers.

Joseph Poore, of the University of Oxford, led a study that consolidated a huge amount of environmental data on 38,700 farms and 1,600 food processors in 119 countries and covered 40 different food products. Poore summarized the upshot of all this research thus: "A vegan diet is probably the single biggest way to reduce your impact on planet Earth, not just greenhouse gases, but global acidification, eutrophication, land use and water use," he told *The Guardian*, adding that the impact of going vegan is far greater than cutting down on flying, or buying an electric car, because these only reduce greenhouse gas emissions. Poore doesn't see "sustainable" animal agriculture as the solution: "Really it is animal products that are responsible for so much of this. Avoiding consumption of animal products delivers far better environmental benefits than trying to purchase sustainable meat and dairy."[29]

DRAWING THE LINE IN OUR EVERYDAY LIVES

IN CHAPTER 1, WE SAW THAT THERE IS STRONG EVIDENCE FOR A capacity to feel pain not only in vertebrates, but also in octopuses and decapod crustaceans. They too are likely to suffer slow and painful deaths when taken from the oceans and killed, and if they are intensively farmed, they may be suffering for their entire lives. So for the sake of these aquatic animals and for the ocean ecology, as well as for impoverished humans, we should avoid eating fish, octopus, lobster, crayfish, and crabs. Beyond these animals, however, we become increasingly uncertain about whether animals commonly eaten can feel pain. We therefore need to ask: If we want to avoid participating in the exploitation of sentient beings, where should we draw the line?

Consider shrimps. In popular usage, that term covers crustaceans of different orders—some are decapods, others are not. But given the extremely large number of shrimps produced in aquaculture—an estimated 440 billion each year—there is sufficient reason to be concerned. In addition, many shrimp hatcheries routinely crush or cut off one or both eyestalks of female shrimps used for breeding. This makes them mature faster, but inflicting this injury—which, if the shrimp is capable of feeling pain, is likely to be painful—has been shown to shorten their lives and lead them to produce offspring more susceptible to stress. So it may be a completely unnecessary mutilation.[30] Shrimp farming is also environmentally damaging and we should not support it.[31]

There are millions of species of insects, and although we cannot say with confidence that any of them can suffer, we also cannot exclude that possibility.[32] When we can give them the benefit of the doubt, at modest cost to ourselves, we should do so. That doesn't mean that it is wrong to slap a mosquito, for as we saw in chapter 1, the capacity to feel pain does not give a being a right to life, but it does mean that if we need to get rid of insects, we should kill them quickly. Flypaper is coated with a sticky substance that is attractive to flies. They become trapped in it and may take hours to die. If there is even a small probability that they are suffering, we should use other methods of getting rid of them.

Commercial insect production raises further ethical questions. Each year more than a *trillion* insects are raised commercially, with the number alive in commercial production on any given day in the range of 79 to 94 billion.[33] In the United States, Canada, and Europe, most of this production is turned into animal feed, but in China and Thailand they are also produced as food for humans. One major reason behind the push to develop insects as food is that, per unit of protein produced, they release only 1 percent of the greenhouse gases of cows raised for beef, and require only about 10 percent of the land, water, and feed. If we knew that insects are not sentient beings, then we could hope that eating insects would become more popular and reduce the demand for factory farmed products from vertebrates. Unfortunately, we don't know that, and in any case, when insects are produced to be fed to animals, the insect production will be propping up the continued exploitation of chickens, pigs, and cows.

There is now a *Journal of Insects as Food and Feed.* Arnold van Huis, the editor and a professor of entomology in the Netherlands, has editorialized on concerns about the welfare of insects in commercial production, concluding that it is possible that insects are sentient beings and recommending that they be treated as such.[34] How likely it is that they are sentient beings, and what treating them as such would involve, will depend on the species of insect raised. The three main species in commercial production are crickets, mealworms, and black soldier fly larvae. Of these, crickets, who have a similar number of neurons as bees, would seem to be the most likely to be capable of suffering.[35] Given that there are around 400 billion crickets produced each year, and between 34 and 41 billion alive on commercial farms at any one time, we should be thinking about how to minimize suffering that they might be able to experience—and we should also be finding better options for feeding ourselves.

In the first edition of this book, when discussing what to eat and what not to eat, I wrote: "Somewhere between a shrimp and an oyster seems as good a place to draw the line as any, and better than most." But several people who liked the book as a whole urged me to reconsider, saying that if I eat oysters, I'm not even a vegetarian, let alone a vegan. The label we use to describe ourselves, however, isn't important. When meat grown from cultured animal cells comes on the market, I will eat it, if only to satisfy my curiosity. The fact that it will be meat doesn't show that we shouldn't eat it. What really matters is whether we are contributing to avoidable suffering, human or nonhuman, or to damaging the environment.

The stronger argument against eating oysters is that we should give them the benefit of the doubt. "We don't need to eat them," one vegan friend said, "and we can't be sure that they don't feel pain." These thoughts led me to say, in the 1990 edition, that it is better to avoid them. Looking at the evidence again for this book, I was pleased to find that Peter Godfrey-Smith, reviewing the latest research, found that the line between a shrimp and an oyster still stands up. For shrimp, he wrote, "recent work has shown surprising complexity that is directly relevant to questions about pain. In oysters, not so."[36]

If, more than forty years ago, I did draw the line in the right place, it was more by good luck than solid research. Then I had argued that organisms evolve a capacity to feel pain because it warns them of dangers from which they can then move away. As oysters don't move, there is no reason why they should be capable of feeling pain. But I also referred to mussels, clams and scallops as if, since they are all bivalves, what applies to oysters will apply to them—an ignorant mistake! To quote Godfrey-Smith again: "Oysters are behaviorally simpler than other mollusks, even other bivalves." Scallops can swim, by pumping water through their shells, and clams can burrow into the sand. Scallops and clams also have eyes, which enable them to sense danger. It remains possible that their movements are mere reflex responses and do not indicate consciousness, but there are stronger grounds for believing that they may be sentient than there are for oysters, which attach themselves to a rock during the larval stage and thereafter remain fixed in one place. As for mussels, here too we cannot generalize, because marine mussels and freshwater mussels are not only different species but in different subclasses of bivalves. Freshwater mussels may use their muscular "feet" to move along the bed of the river or lake they are in, whereas marine mussels—which are the ones I had in mind—attach themselves to something solid and generally stay there. In broad terms, I would still draw the line somewhere between a shrimp and an oyster, but within that gap, I'm finding shades of gray, and I now think that what is true of oysters, and perhaps marine mussels, may not be true of scallops or clams.

Why not avoid the complications and fine-line drawing, and just stop eating any animals at all, giving them all the benefit of the doubt? You can do that, of course, but it doesn't eliminate all risk of causing sentient beings to suffer. Growing vegetables can also cause animals to die painfully: for example, by the methods we use to prevent insects, slugs, snails, birds, and rodents from eating the crops we are growing. More than 97 percent of oysters sold, and about 87 percent of mussels, are farmed, usually grown on ropes or posts in the sea, and that is likely to be less harmful to other animals than growing vegetables is to the land animals just mentioned. No land has to be cleared for them, and they clean the water as they feed, pumping it through their gills to get food. In Chesapeake Bay, for example, oyster aquaculture is being actively encouraged to remove sediment from the water so that underwater grasses can grow and provide habitat for other aquatic animals.[37] So if oysters, and marine mussels are not capable of suffering, eating them may cause *less* animal suffering, and be better for the environment, than eating plants. Close to half of all scallops sold, on the other hand, are dredged up from the seabed, which can destroy entire underwater ecosystems.[38]

FROM THOUGHT TO ACTION

I HAVE DINED WITH PEOPLE WHO READILY ADMIT THAT THEY shouldn't be eating animal products, and especially not factory-farmed animal products; yet they then order chicken. Their values and their food choices are not in harmony, and this discord must get repeated on a daily basis. That can't be a good way to live. Ultimately it is up to each one of us to put our convictions into practice, and a book can't do that for you. But I can try, in the next few pages, to narrow the gap. My aim is to make the transition to plant-based eating easier and more attractive, so that instead of seeing the change of diet as a regrettable moral necessity, you will look forward to a new and interesting cuisine, full of fresh foods as well as unusual dishes from Europe, India, Southeast Asia, China, and the Middle East, dishes so varied as to make the habitual meat, meat, and more meat of most Western diets dull and repetitive by comparison. The enjoyment of such a cuisine is enhanced by the knowledge that its good taste and nourishing qualities were provided with minimal use of arable land and water, low greenhouse gas emissions, and you don't have to worry about complicity in the suffering of the animal whose flesh you are eating.

When Renata, my wife, and I became vegetarians in 1971, we were living in England, where meals were always built around meat. Vegetables were usually boiled until they were soft. After we stopped eating meat, I took an extra delight in fresh vegetables taken straight from the ground and sautéed or stir-fried. I even started growing some of my own vegetables—something I had never thought of doing previously, but that several of my vegetarian friends were also doing. I still do it too, whenever I have the opportunity. Avoiding animal flesh brought me into closer contact with plants, the soil, and the seasons. It also got me into the kitchen. For the first two years of our marriage, Renata had done almost all the cooking because—in accordance with the usual gender-based division of roles—she knew much more about cooking than I did. But given her Eastern European background, most of the main courses she cooked were meat-based, so that knowledge was rendered useless by our decision to stop eating meat. We began exploring new, more vegetarian-friendly cuisines together, and I became the family specialist for Chinese and Indian cooking. In the first edition of this book I had an appendix with some recipes and hints on vegetarian cooking to help readers make the transition to what was, then, still an unusual diet; but by the time I revised the book for the 1990 edition, there were so many excellent vegetarian and vegan cookbooks that the assistance I had provided seemed unnecessary, and I dropped the recipes. To my surprise, several readers asked me to bring the recipes back, so in this edition, I am including just a few of my favorites. You will find them at the end of the book.

In the 1970s, when I spoke in public about the issues discussed in this book, I was often asked what I eat instead of meat. It was clear from the way the question was phrased that the questioner had mentally subtracted the piece of meat from their plate, leaving the mashed potatoes and boiled peas, and was wondering what to put in place of the meat. But my question back to them would be: Why do I need a replacement? Some people find it hard, at first, to change their attitude to a meal. Getting used to meals without a central piece of animal flesh may take time, but once it has happened you will have so many interesting new dishes to choose from that you will wonder why you ever thought it would be difficult to do without flesh foods.

Also in the 1970s, many people were worried about whether they could be adequately nourished without eating meat (let alone being vegan). Now there are so many vegans in public life. Nick Kyrgios, who has a reputation for being the "bad boy" of tennis, says that he is a vegan, not for his health but because he is "trying to be the change I want to see"—and that change is for animals and for the environment.[39] Other vegan athletes include women's soccer star Alex Morgan, basketballers Damian Lillard and DeAndre Jordan, racing driver Lewis Hamilton, women's surfing champion Tia Blanco, ultramarathon runner Scott Jurek, long-distance hiker Josh Garrett (who held the record for hiking the Pacific Crest Trail, from Mexico to Canada), and an entire British soccer team, the Forest Green Rovers. Then, if you need any more evidence that a plant-based diet can give you all the strength, energy, and endurance that anyone needs, there is the remarkable story of Rich Roll, who at the age of forty abandoned an unhealthy lifestyle that included drugs and alcohol, switched to a plant-based diet and began working on his body. Two years later, he was among the top finishers at Ultraman World Championships, and, as an advocate for wellness and plant-based living, became a bestselling author and leading podcaster.

Many people who switch to a plant-based diet say that they feel fitter, healthier, and more zestful than when they ate animal products, and that was my experience too. Nutritional experts no longer dispute about whether eating animal products is essential to good health because the evidence is clear that it is not. The *Lancet*, one of the world's leading medical journals, set up the EAT-Lancet Commission to propose science-based targets for healthy and sustainable diets for a world of 10 billion people. The commission brought together thirty-seven experts in human health, agriculture, political science, and environmental sustainability. On human health, the commission summarized the upshot of two major dietary studies. One, a major North American prospective study, tracked more than 70,000 people on vegan, vegetarian, pescatarian, semi-vegetarian, and omnivore diets for nearly six years and found that the mortality rate of those who ate little or no meat was 12 percent lower than that of the omnivores. Mortality was more sharply reduced in vegans, vegetarians, and pescatarians than in semi-vegetarians, and overall, the reduction was greater in men rather than women.[40] A second study found that eating more plant-based foods was associated with lower rates of type 2 diabetes and coronary heart disease.[41] The commission described these findings as suggesting that "a shift toward a dietary pattern emphasizing whole grains, fruits, vegetables, nuts, and legumes without necessarily becoming a strict vegan, will be beneficial."[42]

In a separate publication, the commission described what it calls a "Planetary Health Diet" that could achieve the desired target of providing a healthy diet for 10 billion people while preserving the environment. This diet should consist, the commission says, of vegetables, fruits, whole grains, plant protein sources, unsaturated plant oils, "and (optionally) modest amounts of animal sources of protein."[43] Note the word "optionally": The commission is stating clearly that we do not need to eat animal products, but it is also not saying that, from a health perspective or an environmental perspective, everyone needs to be strictly vegan. Throughout the report, however, the environmental and health problems of producing and consuming large quantities of animal products are highlighted. It is not my aim to demonstrate that an entirely plant-based diet will enable everyone to be healthier, or live longer, than they would if they ate small quantities of meat or other animal products. To show that the goal of ending speciesism is feasible it is enough to show that, in living in accordance with that goal, we will not be putting our health at risk—and that is what the weight of scientific evidence indicates. It is an extra bonus that, as the studies cited by the EAT-Lancet Commission indicate, for most people living in affluent countries and eating diets high in animal products, eliminating or sharply reducing their intake of animal products will improve their health.

CHAPTER 5

Man's Dominion . . .

a short history of speciesism

To end tyranny we must first understand it.

As a practical matter, the rule of humans over other animals expresses itself in the manner we have seen in chapters 2 and 3 and in related practices like the slaughter of wild animals for sport or for their furs. These practices should not be seen as isolated aberrations. They can be properly understood only as the manifestations of our species' ideology—that is, the attitudes that we, as the dominant animal, have toward the other animals.

In this chapter we shall see how, at different periods in human history, outstanding Western thinkers formulated and defended the attitudes to animals that we have inherited. I concentrate on the Western tradition not because other cultures are inferior—the reverse is often true, so far as attitudes toward animals are concerned—but because it is the tradition I know best. And whether we like it or not, it has, over the past five centuries, spread out from Europe and influenced the rest of the world to a far greater extent than any other intellectual tradition.[1]

Though the material that follows is historical, my aim in presenting it goes beyond a desire to add to your knowledge of the past. When an attitude is so deeply ingrained in our thought that we take it as an unquestioned truth, a challenge to that attitude runs the risk of being ridiculed or dismissed out of hand. It may be possible to shatter the complacency with which the attitude is held by a frontal attack, as I have tried to do in the preceding chapters. An alternative strategy is to attempt to undermine the plausibility of the prevailing attitude by exposing its historical origins. That is what I attempt to do in this chapter. Just as we can now see past beliefs about the inferiority of Blacks and women as self-serving white and male ideologies supporting slavery and the dominance of men over women, so a fresh look at the purported justifications for our exploitation of animals may help us to see these beliefs as spurious ideological camouflages for self-serving practices. We may then be ready to take a more skeptical view of what we have until now regarded as our natural and self-evident entitlement to use animals for our own ends.

The Western belief that our use of animals rests on a sound ethical basis has its roots in two traditions: Judaism, with its belief in our divinely bestowed right of dominion over the animals; and Greek antiquity, with its anthropocentric—or, more specifically, rational being centered—view of the universe. These roots unite in Christianity, and it is through Christianity that they came to prevail in Europe. A more enlightened view of our relations with animals starts to spread only in the eighteenth century, as thinkers begin to take positions that are relatively independent of the Church. Despite the immense progress we have made over the past five centuries in our understanding of the universe and our place in it, when it comes to the way we think about animals, we have still not freed ourselves fully from the influence of beliefs that dominated European thought for more than fifteen hundred years.

We may divide our historical account of the Western tradition into four parts: before Christianity, Christian thought, the Enlightenment, and the modern era.

BEFORE CHRISTIANITY

THE CREATION OF THE UNIVERSE SEEMS A SUITABLE STARTING point. The biblical story of creation sets out very clearly the nature of the relationship between man and animal as the Hebrew people conceived it. It is a superb example of myth echoing reality:

> And God said, Let us make man in our image, after our likeness: and let them have dominion over the fish of the sea, and over the fowl of the air, and over the earth, and over every creeping thing that creepeth upon the earth.[2]

The Bible tells us that God made man in His own image; we may regard this as man making God in his own image. Either way, it allots human beings a special position in the universe as beings that, alone of all living things, are God-like. Moreover, God is explicitly said to have given man dominion over every living thing. It is true that, in the Garden of Eden, this dominion is not said to have involved killing other animals for food. Genesis 1:29 suggests that at first human beings lived off the herbs and fruits of the trees, and Eden has often been pictured as a scene of perfect peace, in which killing of any kind would have been out of place. Man ruled, but in this earthly paradise his was a benevolent despotism.

After the fall of man (for which the Bible holds a woman and an animal responsible), killing animals was clearly permissible. God himself clothed Adam and Eve in animal skins before driving them out of the Garden of Eden. Their son Abel was a keeper of sheep and made offerings of his flock to the Lord. Then came the flood, when the rest of creation was nearly wiped out to punish man for his wickedness. When the waters subsided, Noah thanked God by making burned offerings "of every clean beast, and of every clean fowl." In return, God blessed Noah and gave the final seal to man's dominion:

> And the fear of you and the dread of you shall be upon every beast of the earth, and upon every fowl of the air, upon all that moveth upon the earth, and upon all the fishes of the sea; into your hands are they delivered.
>
> Every moving thing that liveth shall be meat for you; even as the green herb have I given you all things.[3]

This is the basic position of the ancient Hebrew writings toward nonhumans. There is again an intriguing hint that in the original state of innocence we were vegetarian, eating only "the green herb"; but after the fall, the wickedness that followed it, and the flood, we were given permission to add animals to our diet. Beneath the assumption of human dominion that this permission implies, a more compassionate vein of thought still occasionally emerges. For example, the prophet Isaiah condemned animal sacrifices, and the book of Isaiah contains a lovely vision of the time when the wolf will dwell with the lamb, the lion will eat straw like the ox, and "they shall not hurt nor destroy in all my holy mountain." This, however, is a utopian vision, not a command to be followed immediately. Other scattered passages in the Old Testament encourage some degree of kindliness toward animals, such that it is possible to argue that wanton cruelty was prohibited and that "dominion" is really more like a "stewardship," in which we are responsible to God for the care and well-being of those placed under our rule. Nevertheless, there is no serious challenge to the overall view, laid down in Genesis, that the human species is the pinnacle of creation and has God's permission to kill and eat other animals.

The second ancient tradition of Western thought comes out of Greece. Greek thought was not uniform but divided into rival schools, each taking its basic doctrines from some great founder. One of these, Pythagoras, was a vegetarian and encouraged his followers to treat animals with respect, possibly because he believed that when we die our souls migrate to animals. But the most important schools were those of Plato and his pupil Aristotle.

Aristotle's support for slavery is well known; he thought that some men are slaves by nature and that slavery is both right and expedient for them. This fact is essential for understanding his attitude toward animals. Aristotle held that animals exist to serve the purposes of human beings, although, unlike the author of Genesis, he does not drive any deep gulf between human beings and the rest of the animal world. Aristotle does not deny that man is an animal; in fact he defines man as a rational animal. Sharing a common animal nature, however, is not enough to justify equal consideration. For Aristotle, the man who is by nature a slave is undoubtedly a human being and is thus as capable of feeling pleasure and pain as any other human being; yet because he is supposed to be inferior to the free man in his reasoning powers, Aristotle regards him as a "living instrument." Quite openly, Aristotle juxtaposes the two elements in a single sentence: The slave is one who "though remaining a human being, is also an article of property."[4]

If the difference in reasoning powers between human beings is enough to make some men masters and others their property, Aristotle must have thought the right of humans to rule over other animals is too obvious to require much argument. Nature, he held, is essentially a hierarchy in which those with less reasoning ability exist for the sake of those with more:

> Plants exist for the sake of animals, and brute beasts for the sake of man—domestic animals for his use and food, wild ones (or at any rate most of them) for food and other accessories of life, such as clothing and various tools. Since nature makes nothing purposeless or in vain, it is undeniably true that she has made all animals for the sake of man.[5]

It was the views of Aristotle, rather than those of Pythagoras, that were to become part of the later Western tradition.

CHRISTIAN THOUGHT

AS WE SHALL SEE, CHRISTIANITY WAS EVENTUALLY TO UNITE JEWish and Greek ideas about animals. But Christianity was founded and became powerful under the Roman empire, and we can see its initial effect best if we compare Christian attitudes with those they replaced. The Roman empire was built through wars of conquest, and it needed to devote much of its energy and revenue to the military forces that defended and extended its vast territory. These conditions did not foster sentiments of sympathy for the weak; the martial virtues set the tone for society. Within Rome itself, far from the fighting on the frontiers, the character of Roman citizens was supposedly toughened by the so-called games. Although everyone knows how Christians were thrown to the lions in the Colosseum, the significance of the games as an indication of the possible limits of sympathy and compassion of apparently—and in other respects genuinely—civilized people is rarely appreciated. Men and women looked upon the slaughter of both human beings and other animals as a normal source of entertainment; and this continued for centuries with scarcely a protest.

The nineteenth-century historian W. E. H. Lecky describes the development of the Roman games from their beginning as a combat between two gladiators until, to stimulate flagging interest, wild animals were added, and the numbers of animals then increased:

> In a single day, at the dedication of the Colosseum by Titus, five thousand animals perished. Under Trajan, the games continued for one hundred and twenty-three successive days. Lions, tigers, elephants, rhinoceroses, hippopotami, giraffes, bulls, stags, even crocodiles and serpents were employed to give novelty to the spectacle. Nor was any form of human suffering wanting. . . . Ten thousand men fought during the games of Trajan. Nero illumined his gardens during the night by Christians burning in their pitchy shirts. Under Domitian, an army of feeble dwarfs was compelled to fight. . . . So intense was the craving for blood, that a prince was less unpopular if he neglected the distribution of corn than if he neglected the games.[6]

The Romans were not without moral feelings or the capacity for reflecting on an ethical life. Thinkers like Cicero, Seneca, the former slave Epictetus and the emperor Marcus Aurelius show a high regard for justice, public duty, and reflection on how best to live one's life. What the games show, with hideous clarity, is that there was a sharp limit to these moral feelings and reflections. If a being came within this limit, activities comparable to what occurred at the games would have been an intolerable outrage; when a being was outside the sphere of moral concern, however, the infliction of suffering was merely entertaining. Some human beings—slaves, criminals, and military captives especially—and all animals fell outside this sphere.

It is against this background that the impact of Christianity must be assessed. Christianity brought into the Roman world the idea of the unique place in creation of the human species, an idea that it inherited from the Jewish tradition but insisted upon with still greater emphasis because of the importance it placed on the belief that human beings, alone of all beings living on earth, were destined for life after bodily death. With this came the distinctively Christian idea of the sanctity of all human life. There have been religions, especially in the East, that have taught that all life is sacred; and there have been many others that have held it gravely wrong to kill members of one's own social, religious, or ethnic group; but Christianity spread the idea that every human life—and only human life—is sacred. Even the newborn infant and the fetus in the womb have immortal souls, and so their lives are as sacred as those of adults.

In its application to human beings, the new doctrine was progressive, and it led to an enormous expansion of the limited moral sphere of the Romans. So far as other species are concerned, however, this same doctrine served to confirm and further depress the lowly position nonhumans held in the Hebrew scriptures. While these writings asserted human dominion over other species, they did at least show flickers of concern for their sufferings. The Christian scriptures, on the other hand, are completely lacking in any injunction against cruelty to animals, or any recommendation to consider their interests. Jesus himself is described as showing apparent indifference to the fate of nonhumans when he induced two thousand swine to hurl themselves into the sea—an act that was apparently quite unnecessary, since Jesus was well able to cast out devils without inflicting them upon any other creature.[7] St. Paul insisted on reinterpreting the old Mosaic law that forbade muzzling the oxen that trod out the corn: "Doth God care for oxen?" Paul asks scornfully. No, he answered, the law was intended "altogether for our sakes."[8]

Similarly, the examples given by Jesus were used by later influential Christians to reinforce the idea of human dominance. Referring to the incident of the swine and the episode in which Jesus cursed a fig tree, St. Augustine wrote:

> Christ himself shows that to refrain from the killing of animals and the destroying of plants is the height of superstition, for judging that there are no common rights between us and the beasts and trees, he sent the devils into a herd of swine and with a curse withered the tree on which he found no fruit. Surely the swine had not sinned, nor had the tree.

Jesus was, according to Augustine, trying to show us that we need not govern our behavior toward animals by the moral rules that govern our behavior toward humans. That is why he transferred the devils to swine instead of destroying the devils, as he could easily have done.[9] On this basis the outcome of the interaction of Christian and Roman attitudes is not difficult to guess. It can be seen most clearly by looking at what happened to the Roman games after the conversion of the empire to Christianity. Christian teaching was implacably opposed to gladiatorial combats. The gladiator who survived by killing his opponent was regarded as a murderer. Mere attendance at these combats made the Christian liable to excommunication, and by the end of the fourth century combats between human beings had been suppressed altogether. On the other hand, the moral status of killing or torturing any nonhuman remained unchanged. Combats with wild animals continued into the Christian era, and apparently decreased only because the declining wealth and extent of the empire made wild animals more difficult to obtain. Indeed, these combats may still be seen, in the modern form of the bullfight, in Spain, France, Portugal, Mexico, Colombia, Venezuela, Peru, and Ecuador.

What is true of the Roman games is also true more generally. Christianity left nonhumans as decidedly outside the pale of sympathy as they ever were in Roman times. Consequently, while attitudes toward human beings were softened and improved beyond recognition, attitudes toward other animals remained as callous and brutal as they were in Roman times. Indeed, not only did Christianity fail to temper the worst of Roman attitudes toward other animals; it unfortunately succeeded in extinguishing for a long, long time the spark of a wider compassion that had been kept alight by a tiny number of more gentle people.

There had been just a few Romans who had shown compassion for suffering, irrespective of the species of the being who suffered. Some even expressed repulsion at the use of sentient creatures for human pleasure, whether at the gourmet's table or in the arena. Ovid, Seneca, Porphyry, and Plutarch all wrote along these lines. Plutarch has the honor, according to Lecky, of being the first to advocate strongly the kind treatment of animals on the ground of universal benevolence, independently of any belief in the transmigration of souls. Nor should we omit Apuleius, the North African author of *The Golden Ass*, an entertaining and bawdy tale of a man turned into donkey that is arguably both the world's oldest surviving novel and the first work to portray the world from the perspective of an animal, including the many different forms of cruelty that were then inflicted on donkeys. We have to wait nearly sixteen hundred years before any Christian writer attacks cruelty to animals with similar emphasis and detail to Plutarch, on any ground other than that it may encourage a tendency toward cruelty to humans; and it is not until the late nineteenth century that we have, in Anna Sewell's *Black Beauty*, a novel that takes an animal's perspective as thoroughly as *The Golden Ass*.[10]

Some early Christians expressed concern for animals. There is a prayer written by St. Basil of Caesarea that urges kindness toward animals, a remark by St. John Chrysostom to the same effect, and a teaching of St. Isaac the Syrian. There were even saints like Neot, who sabotaged hunts by rescuing stags and hares from the hunters.[11] In later periods there were many humane Catholics who did their best to ameliorate the position of their Church with regard to animals. Yet most remained limited by the essentially speciesist outlook of their religion. The most famous of them all, St. Francis of Assisi, illustrates this well. He is the outstanding exception to the rule that Catholicism discourages concern for the welfare of nonhuman beings. "If I could only be presented to the emperor," he is reported as saying, "I would pray him, for the love of God, and of me, to issue an edict prohibiting anyone from catching or imprisoning my sisters the larks, and ordering that all who have oxen or asses should at Christmas feed them particularly well." Many legends tell of his compassion, and the story of how he preached to the birds certainly seems to imply that the gap between them and humans was less than other Christians supposed—even if, as has been suggested, he preached to the birds in order to indicate to his human audiences that they were not attending to his teachings any better than the birds did. To look only at Francis's attitude toward larks and other animals, however, gives a misleading impression of his thinking. It was not only sentient creatures whom St. Francis addressed as his sisters;

the sun, the moon, wind, fire, all were brothers and sisters to him. His contemporaries described him as taking "inward and outward delight in almost every creature, and when he handled or looked at them his spirit seemed to be in heaven rather than on earth." This delight extended to water, rocks, flowers, and trees. This is an account of a person in a religious ecstasy, deeply moved by a feeling of oneness with all of nature. People from a variety of religious and mystical traditions appear to have had such experiences and have expressed similar feelings of universal love. Seeing Francis in this light makes the breadth of his love and compassion more readily comprehensible. It also enables us to see how his love for all creatures could coexist with a theological position that was quite orthodox in its speciesism.

This kind of ecstatic universal love can be a powerful source of compassion and goodness, but the lack of rational reflection can also do much to counteract its beneficial consequences. If we love rocks, trees, plants, larks, and oxen equally, we may lose sight of the essential differences between them—most important, the fact that some are capable of feeling pain and others are not. We may then think that since we have to eat to survive, and since we cannot eat without killing something we love, it does not matter which we kill. Perhaps it was for this reason that St. Francis's love for birds and oxen appears not to have led him to cease eating them. The rules he drew up for the conduct of the friars in the order he founded did not forbid eating meat, and when Francis returned from long travels in 1220 and found that those he had left in charge had instituted such a prohibition, he reversed it.[12]

Francis is, for Catholics, the patron saint of animals, and on his feast day, October 4, many churches invite parishioners to bring their animals to be blessed. But neither Francis nor any of the other prominent early Christian thinkers concerned about animals has managed to divert mainstream Christian thinking from its exclusively speciesist preoccupation. To demonstrate this lack of influence, consider the views of the most influential of them, St. Thomas Aquinas.

Aquinas was born in 1225, the year before Francis died. His major project, the *Summa theologica*, attempted to sum up theological knowledge and reconcile it with the worldly wisdom of the philosophers, which in effect meant Aristotle, whom Aquinas refers to simply as "the Philosopher." If any single philosophy may be taken as representative of Christian philosophy prior to the Reformation, and of Roman Catholic philosophy almost to the present day, it is the combination of Aristotelian philosophy and Christian theology developed by Thomas Aquinas and known as Thomism.

We may begin our investigation of Thomistic ethics regarding animals by asking whether the Christian prohibition on killing applies to creatures other than humans and, if not, why not. Aquinas answers:

> There is no sin in using a thing for the purpose for which it is. Now the order of things is such that the imperfect are for the perfect. Things, like plants which merely have life, are all alike for animals, and all animals are for man. Wherefore it is not unlawful if men use plants for the good of animals, and animals for the good of man, as the Philosopher states (*Politics* I, 3).
>
> Now the most necessary use would seem to consist in the fact that animals use plants, and men use animals, for food, and this cannot be done unless these be deprived of life, wherefore it is lawful both to take life from plants for the use of animals, and from animals for the use of men.

> In fact this is in keeping with the commandment of God himself (Gen. 1:29–30 and Gen. 9:3).[13]

It is odd that Aquinas suggests that it is necessary for humans to eat animals, for he was familiar with Augustine's writings against Manichaeism, a rival religion to Christianity in Augustine's time, and one of the points of difference between Christianity and Manichaeism that Augustine used against them was that they forbade the killing and eating of animals. (It was with the Manichaens in mind that Augustine explains Jesus's drowning of the pigs as intended to teach us that to refrain from killing animals is "the height of superstition.")

The major ethical claim Aquinas makes in this passage, however, is his acceptance of the Aristotelian view that the "imperfect" exist for the "more perfect" and that we humans are, of course, the most perfect of all. Hence we are entitled to kill to sustain our lives, but animals who kill human beings for food are in a quite different category:

> Savagery and brutality take their names from a likeness to wild beasts. For animals of this kind attack man that they may feed on his body, and not for some motive of justice, the consideration of which belongs to reason alone.[14]

Human beings, of course, would not kill for food unless they had first considered the justice of so doing!

So human beings may kill other animals and use them for food; but are there perhaps other things that we may not do to them? Is the suffering of other creatures in itself an evil? If so, would it not for that reason be wrong to make them suffer, or at least to make them suffer unnecessarily?

Aquinas does not say that cruelty to "irrational animals" is wrong in itself. He has no room for wrongs of this kind in his moral schema, for he divides sins into those against God, those against oneself, and those against one's neighbor. The limits of morality thus once again exclude nonhumans. There is no category for sins against them.[15]

Although it is not a sin to be cruel to nonhumans, is it perhaps charitable to be kind to them? No, Aquinas explicitly excludes this possibility as well. Charity, he says, does not extend to irrational creatures for three reasons: They are "not competent, properly speaking, to possess good, this being proper to rational creatures"; we have no fellow-feeling with them; and, finally, "charity is based on the fellowship of everlasting happiness, to which the irrational creature cannot attain." It is only possible to love these creatures, we are told, "if we regard them as the good things that we desire for others"—that is, "to God's honor and man's use." In other words, we cannot lovingly give food to animals because they are hungry, but we can if we think of them as someone's dinner.[16]

All this might lead us to suspect that Aquinas simply doesn't believe that animals other than human beings are capable of suffering at all. This view has been held by other philosophers, and for all its apparent absurdity, to attribute it to Aquinas would at least excuse him of the charge of indifference to suffering. This interpretation, however, is ruled out by his own words. In the course of a discussion of some of the mild injunctions against cruelty to animals in the Hebrew scriptures, Aquinas proposes that we distinguish reason and passion. So far as reason is concerned, he tells us:

> It matters not how man behaves to animals, because God has subjected all things to man's power and it is in this sense that the Apostle says that God has no care for oxen,

> because God does not ask of man what he does with oxen or other animals.

On the other hand, where passion is concerned, our pity is aroused by animals, because "even irrational animals are sensible to pain"; nevertheless, Aquinas regards the pain that animals suffer as insufficient reason to justify the scriptural injunctions, and therefore:

> Now it is evident that if a man practice a pitiable affection for animals, he is all the more disposed to take pity on his fellow-men, wherefore it is written (Proverbs 12:10) "The just regardeth the life of his beast."[17]

So Aquinas arrives at the oft-repeated view that the only reason against cruelty to animals is that it may lead to cruelty to human beings. No argument could reveal the essence of speciesism more clearly.

The Italian Renaissance, boosted by an influx of scholars escaping the collapse of the Byzantine empire and the Ottoman takeover of Constantinople in 1453, brought new humanist ways of thinking, in opposition to medieval scholasticism. But as far as attitudes toward animals are concerned, Renaissance humanism was, after all, humanism; and the meaning of this term has nothing to do with humanitarianism, the tendency to act humanely. Instead, the central feature of Renaissance humanism is its insistence on the value and dignity of human beings and on our central place in the universe. "Man is the measure of all things," a phrase revived in Renaissance times from the ancient Greeks, is the theme of the period.

Instead of a depressing concentration on the sinfulness and weakness of human beings in comparison to the infinite power of

God, Renaissance humanists emphasized the uniqueness of human beings, their free will, their potential, and their dignity; but to do so, they contrasted human powers with the limited nature of the "lower animals." Renaissance thinkers wrote self-indulgent essays in which they said that "nothing in the world can be found that is more worthy of admiration than man" and described humans as "the center of nature, the middle of the universe, the chain of the world."[18] If the Renaissance marks in some respect the beginning of modern thought, this does not apply to attitudes toward animals.

Around this time, however, we do find the first genuine dissenters: Leonardo da Vinci was teased by his friends for being so concerned about the sufferings of animals that he stopped eating them;[19] and Giordano Bruno, influenced by the new Copernican astronomy, which allowed for the possibility of other planets, some of which could be inhabited, ventured to assert that "man is no more than an ant in the presence of the infinite." Bruno was burned at the stake in 1600 for refusing to recant his heresies.

Michel de Montaigne's favorite author was Plutarch, and his attack on the humanist assumptions of his age would have met with the approval of that gentle Greek:

> Presumption is our natural and original disease. . . .'Tis by the same vanity of imagination that [man] equals himself to God, attributes to himself divine qualities, and withdraws and separates himself from the crowd of other creatures.[20]

It is surely not a coincidence that the writer who rejects such self-exaltation should also, in his essay "On Cruelty," be among the first writers since Roman times to assert that cruelty to animals is wrong in itself, quite apart from its tendency to lead to cruelty to human beings.

At this time, the ancient conceptions of the universe and the central place of humans in it, whether derived from the Judeo-Christian religious tradition or from Aristotle, were under great pressure from the scientific discoveries made by scientists like Copernicus and Galileo. One might think, therefore, that the status of nonhumans would now begin to improve. But the absolute nadir was still to come.

The last, most bizarre, and—for the animals—most painful outcome of Christian doctrines emerged in the first half of the seventeenth century, in the philosophy of René Descartes. In some respects, Descartes was a distinctively modern thinker. His apparent openness to doubting everything gives him a good claim to the title of "father of modern philosophy," and he was also a pioneer of analytic geometry, in which a good deal of modern mathematics has its origins. Yet in some ways this conception of Descartes is misleading. He was open to doubt, but he remained a Christian, and his beliefs about animals arose from an attempt to combine his support for science and his commitment to Christian doctrines.

Under the influence of the new and exciting science of mechanics, Descartes held that everything that consisted of matter was governed by mechanistic principles, like those that explained the operation of a clock. But what about us, then? The human body is composed of matter and is part of the physical universe. Are we also machines? Is our behavior determined by the laws of science?

Descartes was able to escape the unpalatable and heretical view that humans are machines by bringing in the idea of the soul. There are, Descartes said, not one but two kinds of things in the universe: things of the spirit or soul, as well as things of a physical or material nature. Human beings are conscious, and consciousness cannot have its origin in matter. Descartes identified consciousness with the immortal soul, specially created by God, which survives the decomposition of the physical body. In conformity with Christian teachings, Descartes said that human beings are unique in having both a material body and a conscious soul. (Angels and other immaterial beings exist, in this view, as pure consciousness.) Thus in the philosophy of Descartes, the Christian doctrine that animals do not have immortal souls has the extraordinary consequence that they do not have consciousness either. They are, he said, mere machines, automata. They experience neither pleasure nor pain, nor anything else. Although they may squeal when cut with a knife, or writhe in their efforts to escape contact with a hot iron, this does not, Descartes said, mean that they feel pain in these situations. They are governed by the same principles as a clock, and if their actions are more complex than those of a clock, it is because the clock is a machine made by humans, while animals are infinitely more complex machines, made by God.[21]

This "solution" to the problem of locating consciousness in a materialistic world seems paradoxical to us, as it did to many of Descartes's contemporaries, but at the time it was also thought to have important advantages. It provided a reason for believing in a life after death, something that Descartes thought was "of great importance," since "the idea that the souls of animals are of the same nature as our own, and that we have no more to fear or to hope for after this life than have the flies and ants" was an error that was apt to lead to immoral conduct.[22] It also eliminated a difficult aspect of the ancient theological problem of evil: In a world created by an omnipotent, omniscient, and omnibenevolent God, why would animals suffer? After all, humans suffer because they have sinned—or at least inherited original sin from Adam and Eve—and they will be recompensed for their suffering in an afterlife. But animals are innocent, they are not descended from Adam and Eve, and they have no afterlife. Descartes's answer to this otherwise apparently insuperable objection to the existence of God, as traditionally conceived by Christians, is breathtakingly simple: Animals don't suffer.[23]

Descartes was also aware of a very practical advantage to the denial of animal suffering. It is, he claimed, "not so much cruel to animals as indulgent to men—at least to those who are not given to the superstitions of Pythagoras—since it absolves them from the suspicion of crime when they eat or kill animals."[24]

For Descartes the scientist, the doctrine had still another fortunate result. The rise of a more scientific approach to biology and medicine meant that the practice of dissecting live animals was then becoming widespread in Europe. Since there were no anesthetics, these experiments must have caused the animals to behave in a way that would indicate, to most of us, that they were suffering extreme pain. Descartes's theory allowed the experimenters to dismiss any qualms they might feel under these circumstances. Descartes himself dissected living animals in order to advance his knowledge of anatomy, and many of the leading physiologists of the period declared themselves Cartesians and mechanists. The following eyewitness account of some of these experimenters, working at the Jansenist seminary of Port Royal in the late seventeenth century, makes clear the convenience of Descartes's theory:

> They administered beatings to dogs with perfect indifference, and made fun of those who pitied the creatures as if they felt pain. They said the animals were clocks; that the cries they emitted when struck were only the noise of

> a little spring that had been touched, but that the whole body was without feeling. They nailed poor animals up on boards by their four paws to vivisect them and see the circulation of the blood which was a great subject of conversation.[25]

From this point, it really was true that the status of animals could only improve.

THE ENLIGHTENMENT

DAVID HUME, THE GREATEST PHILOSOPHER OF THE SCOTTISH ENlightenment, addressed the issue of our treatment of animals by saying that we are "bound by the laws of humanity to give gentle usage to these creatures."[26] "Gentle usage" is a phrase that nicely sums up the kinder attitude that began to spread in the eighteenth century: We are entitled to use animals, but we ought to do so gently. The tendency of the age was toward greater refinement and civility, more benevolence and less brutality, and animals as well as humans benefited from this tendency. The eighteenth century was also the period in which we rediscovered "Nature": Jean-Jacques Rousseau's noble savage, strolling naked through the woods, picking fruits and nuts as he went, was the culmination of this idealization of nature. By seeing ourselves as part of nature, we regained a sense of kinship with "the beasts."

The new vogue for experimenting on animals may itself have been partly responsible for a change in attitudes toward animals, for the experiments revealed a remarkable similarity between the physiology of human beings and other animals. Strictly, this was not inconsistent with what Descartes had said, but it made his views less plausible. Voltaire put it well:

> There are barbarians who seize this dog, who so greatly surpasses man in fidelity and friendship, and nail him down to a table and dissect him alive, to show you the mesaraic veins! You discover in him all the same organs of feeling as in yourself. Answer me, mechanist, has Nature arranged all the springs of feeling in this animal to the end that he might not feel?[27]

In England, Alexander Pope opposed the practice of cutting open fully conscious dogs by arguing that although "the inferior creation" has been "submitted to our power," we are answerable for the "mismanagement" of it.[28] As the thought that we are the managers of "the inferior creation" shows, this era was still very far from breaking with the dominant view of our right to rule over, and use, the animals. At best, man was seen in the role of benevolent father of the family of animals.

In France, the growth of anticlerical feeling was favorable to the status of animals. Voltaire, who delighted in fighting dogmas of all kinds, compared Christian practices unfavorably with those of the Hindu. He went further than the contemporary English advocates of kind treatment when he referred to the barbarous custom of supporting ourselves upon the flesh and blood of beings like ourselves"—although apparently he continued to practice this particular barbarity himself.[29] Rousseau also seems to have recognized the strength of the arguments for vegetarianism without actually adopting the practice; his treatise on education, *Emile*, contains a long and mostly irrelevant passage from Plutarch that attacks the use of animals for food as unnatural, unnecessary, bloody murder.[30]

The Enlightenment did not affect all thinkers equally in their attitudes toward animals. Immanuel Kant, in his lectures on ethics, told his students: "So far as animals are concerned, we have no direct duties. Animals are not self-conscious, and are there merely as a means to an end. That end is man."[31] But in the same year that Kant gave these lectures—1780—Jeremy Bentham completed his *Introduction to the Principles of Morals and Legislation*. There, in a passage I have already quoted in the first chapter of this book, he gave the definitive answer to Kant: "The question is not, Can they reason? nor, Can they talk? but, Can they suffer?" In comparing the position of animals with that of African slaves—and looking forward to the day "when the rest of the animal creation may acquire those rights which never could have been withholden from them but by the hand of tyranny"—Bentham may have been the first to denounce "man's dominion" as tyranny rather than legitimate government.

The intellectual progress made in the eighteenth century was followed, in the nineteenth century, by some practical improvements in the conditions of animals in the form of laws against wanton cruelty to animals. The first debate over legal rights for animals took place in the British Parliament, and it indicated that Bentham's ideas were yet to have a significant influence on its members. Under discussion was a bill to prohibit the "sport" of bull-baiting, introduced into the House of Commons in 1800. George Canning, the foreign secretary, described it as "absurd" and asked rhetorically: "What could be more innocent than bull-baiting, boxing, or dancing?" Since no attempt was being made to prohibit boxing or dancing, it appears that this astute statesman had missed the point of the bill he was opposing—he thought it an attempt to outlaw gatherings of "the rabble" that might lead to immoral conduct.[32] The presupposition that made this mistake possible was that conduct that injures only an animal cannot possibly be worth legislating about—a presupposition shared by the *Times*, which devoted an editorial to the principle that "whatever meddles with the private personal disposition of man's time or property is tyranny. Till another person is injured there is no room for power to interpose." The bill was defeated.

In 1821 Richard Martin, an Irish gentleman landowner who represented Galway in the House of Commons, proposed a law to prevent the ill treatment of horses. The following account conveys the tone of the ensuing debate:

> When Alderman C. Smith suggested that protection should be given to asses, there were such howls of laughter that the *Times* reporter could hear little of what was said. When the Chairman repeated this proposal, the laughter was intensified. Another member said Martin would be legislating for dogs next, which caused a further roar of mirth, and a cry "And cats!" sent the House into convulsions.[33]

This bill failed too, but in the following year Martin succeeded with a bill that made it an offense to "wantonly" mistreat certain domestic animals, provided that they were "the property of any other person or persons." For the first time, cruelty to animals was a punishable offense. Despite the mirth of the previous year, asses were included; dogs and cats, however, were still beyond the pale. More significantly, Martin had had to frame his bill so that it resembled a measure to protect items of private property, for the benefit of the owner, rather than for the sake of the animals themselves.[34]

The bill was now law, but it still had to be enforced. Since the victims could not make a complaint, Martin and a number of other notable humanitarians formed a society to gather evidence and bring prosecutions. So began the world's first animal welfare organization, the Society for the Prevention of Cruelty to Animals, later to receive the blessing of Queen Victoria and add the prefix "Royal."

Toward the end of the 1830s, Charles Darwin wrote in one of his notebooks an idea that was then emerging from his research: "Man in his arrogance thinks himself a great work, worthy of the interposition of a deity. More humble and, I believe, true, to consider him created from animals."[35] Another twenty years were to pass before, in 1859, Darwin considered that he had accumulated enough evidence to publish his theory. Even then, in *On the Origin of Species*, Darwin avoided any discussion of the extent to which his theory of the evolution of one species from another could be applied to humans, saying only that the work would illuminate "the origin of man and his history." In fact, Darwin already had extensive notes on the theory that *Homo sapiens* had descended from other animals, but he decided that publishing this material would "only add to the prejudices against my views."[36] Only in 1871, when many scientists had accepted the general theory of evolution, did Darwin publish *The Descent of Man*, thus making explicit what had been concealed in a single sentence of his earlier work.

Thus began a revolution in human understanding of the relationship between ourselves and the nonhuman animals . . . or, rather, that's how things would have developed if the fact that humans are "created from animals" had not remained merely an intellectual conviction but had penetrated sufficiently deeply into our psyche to make repugnant the very thought of killing and eating our fellow animals. Intellectually the Darwinian revolution was genuinely revolutionary. The fact of our common evolution with other animals overthrows practically every earlier justification of our supreme place in creation and our dominion over the animals. We now know that we are not the special creation of God, made in the divine image and set apart from the animals. Darwin himself insisted that the differences between humans and animals are not so great as was generally supposed. Chapter 3 of *The Descent of Man* is devoted to a comparison of the mental powers of humans and the "lower animals," and Darwin summarizes the results of this comparison by saying that "the various emotions and faculties, such as love, memory, attention and curiosity, imitation, reason etc., of which man boasts, may be found in an incipient, or even sometimes in a well-developed condition, in the lower animals." The fourth chapter of the same work goes still further, affirming that our own moral sense can also be traced back to social instincts in animals that lead them to take pleasure in each other's company, feel sympathy for each other, and perform services of mutual assistance.[37]

The storm of resistance that met the idea that we are descended from animals—a story too well known to need retelling here—is an indication of the extent to which speciesist myths had come to dominate Western thought. With the eventual acceptance of Darwin's theory, we reach a modern understanding of nature, one that has since changed in detail but not fundamentally. Only those who prefer religious faith to beliefs based on reasoning and evidence can still maintain that the human species is the special darling of the entire universe, or that other animals were created to provide us with food, or that we have divine authority over them, and divine permission to kill them.

When we add this intellectual revolution to the growth of humanitarian feeling that preceded it, we might think that all will now be well. Yet, as I hope the preceding chapters have made plain, the human "hand of tyranny" is still clamped down on other species, and we probably inflict more pain on animals now than at any other time in history. What went wrong?

THE MODERN ERA

THE ERA THAT BEGAN WITH BENTHAM CAN BEST BE CHARACTERized, with respect to what progressive thinkers say about animals, as the era of excuses for eating meat. The excuses used vary, and some of them are still around today, so they are worth examining.

First there is the Divine Excuse, illustrated by William Paley's widely read *Principles of Moral and Political Philosophy*, first published in 1785 but taught at Cambridge University until the middle of the nineteenth century. In setting out "the General Rights of Mankind," Paley first says that "some excuse seems necessary" for the pain and loss of life we cause animals. Then he considers whether the fact that they prey on each other gives us a sufficient justification for killing them. He answers that it does not, because they need to kill to survive, whereas we can live on "fruits, pulses, herbs and roots," as, Paley knows, many people in India do. He then concludes that it would be difficult to support our right to eat animals by any nonreligious arguments, and so to justify this practice, we have no alternative but to rely on the permission given us in Genesis.[38]

Paley is only one of many who have appealed to revelation when they found themselves unable to give a rational justification for a diet consisting of other animals. Francis Wayland, whose *Elements of Moral Science*, first published in 1835 and perhaps the most widely used work on moral philosophy in nineteenth-century America, did the same.[39] Henry S. Salt, in his autobiography, *Seventy Years among Savages* (an account of his life in England), records a conversation he had when he was a master at Eton College. He had recently become a vegetarian; now for the first time he was to discuss his practice with a colleague, a distinguished science teacher. With some trepidation he awaited the verdict of the scientific mind on his new beliefs; when it came, it was: "But don't you think that animals were sent to us for food?"[40]

Benjamin Franklin was for some years a vegetarian until, as he recounts in his *Autobiography,* he was watching some friends fishing and noticed that some of the fish they caught had eaten other fish. He therefore concluded, "If you eat one another, I don't see why we may not eat you." Franklin, however, was at least honest enough to admit that he reached this conclusion only after the fish was in the frying-pan and had begun to smell "admirably well." He adds that one of the advantages of being a "reasonable creature" is that one can find a reason for whatever one wants to do.[41]

It is also possible for a deep thinker to avoid confronting the troublesome issue of animal suffering by regarding it as altogether too profound for the human mind to comprehend. As Thomas Arnold, the reforming English educator and headmaster of Rugby School, wrote: "The whole subject of the brute creation is to me one of such painful mystery that I dare not approach it."[42]

This attitude was shared by the French historian Jules Michelet, who expressed it more dramatically:

> Animal Life, somber mystery! Immense world of thoughts and of dumb sufferings. All nature protests against the barbarity of man, who misapprehends, who humiliates, who tortures his inferior brethren. Life, death! The daily murder which feeding upon animals implies—those hard and bitter problems sternly placed themselves before my mind. Miserable contradiction. Let us hope that there may be another sphere in which the base, the cruel fatalities of this may be spared to us.[43]

Michelet seems to have believed that we cannot live without killing; if so, his anguish at this "miserable contradiction" must have been in inverse proportion to the amount of time he gave to examining it.

Another to accept the comfortable error that we must kill to live was the German philosopher Arthur Schopenhauer, who was influential in introducing Eastern ideas to Europe. In sharp and scornful prose, he contrasted the "revoltingly crude" attitudes toward animals prevalent in Western philosophy and religion with those of Buddhists and Hindus. Many of his acute criticisms of Western attitudes are still appropriate today. After one particularly biting passage, however, Schopenhauer briefly considers the question of killing for food. He can hardly deny that human beings can live without killing—he knows too much about India for that—but he claims that "without animal food the human race could not even exist in the North." He adds that the death of the animal should be made "even easier" by means of chloroform, but he gives no basis for the geographical distinction he draws.[44]

Even Bentham flinched on the issue of eating animals, for in the same passage in which he said that the question is not whether animals can reason or talk but whether they can suffer, he also wrote:

> There is very good reason why we should be suffered to eat such of them as we like to eat; we are the better for it, and they are never the worse. They have none of those long-protracted anticipations of future misery which we have. The death they suffer in our hands commonly is, and always may be, a speedier, and by that means a less painful one, than that which would await them in the inevitable course of nature.

One cannot help feeling that in these passages Schopenhauer and Bentham lowered their normal standards of argument. Their support for humane methods of slaughter is laudable, but neither Schopenhauer nor Bentham considers the suffering involved in rearing and slaughtering animals on a commercial basis. At the time when Schopenhauer and Bentham wrote, slaughter was an even more horrific affair than it is today. Animals were forced to cover long distances on foot, driven to slaughter by drovers who had no concern but to complete the journey as quickly as possible; they might spend two or three days in the slaughter yards, without food, perhaps without water; they were then slaughtered by barbaric methods, without any form of prior stunning.[45] They may not have had a long-protracted anticipation of future misery, but they are likely to have had a sense of danger and been fearful or distressed from the time they entered the slaughter yard and smelled the blood of their fellows. Bentham and Schopenhauer would not, of course, have approved of this, yet they continued to support the process by consuming its products and justifying the general practice of which it was part. In this respect Paley seems to have had a more accurate conception of what was involved in eating flesh. He, however, could safely look the facts in the face because he believed that he had divine permission to continue to eat animals; Schopenhauer and Bentham were too rational to avail themselves of this excuse and so had to turn their gaze away from the ugly reality.

Darwin also retained the moral attitudes toward animals of earlier generations, continuing to dine on the flesh of those who, as he had described with numerous compelling examples, are capable of love, memory, curiosity, reason, and sympathy for each other. He even refused to sign a petition urging the RSPCA to press for legislative control (not abolition) of experiments on animals.[46] His followers went out of their way to emphasize that although our origins are natural rather than divine, our understanding of evolution does not lead to any change in our status. In reply to the accusation that Darwin's ideas undermined the dignity of man, T. H. Huxley, Darwin's greatest champion, said: "No-one is more strongly convinced than I am of the vastness of the gulf between civilized man and the brutes; our reverence for the nobility of mankind will not be lessened by the knowledge that man is, in substance and in structure, one with the brutes."[47]

It is a distinctive characteristic of an ideology that it resists refutation. If the foundations of an ideological position are knocked out from under it, new foundations will be found, or else the ideological position will just hang there, defying the logical equivalent of the laws of gravity. In the case of attitudes toward animals, the latter seems to have happened, as illustrated by Huxley: He knows perfectly well that the old reasons for assuming a vast gulf between "man" and "the brutes" no longer stand up, but he continues to believe in the existence of such a gulf nevertheless. (Or at least he believes in the vast gulf between "civilized man" and animals. He accepts the racist attitudes that were common in his day and seems willing to grant that the gulf between the animals and humans who are not "civilized" is not so vast.)

Although the modern view of our place in the world differs enormously from all the earlier views we studied, in the practical matter of how we act toward other animals, little has changed. If animals are no longer entirely outside the moral sphere, they are still separated from humans, in a special section near the outer rim. Their interests are allowed to count only when they do not clash with human interests. If there is a clash—even a clash between a lifetime of suffering for an animal and the gastronomic preferences of a human being—the interests of the nonhuman are disregarded. The moral attitudes of the past are too deeply embedded in our thought and our practices to be upset by a mere change in our knowledge of ourselves and of other animals.

Some ideologies are especially slow to change. For seven

centuries, the Roman Catholic Church continued to follow the teachings of Thomas Aquinas on animals. In the middle of the nineteenth century, Pope Pius IX refused to allow a Society for the Prevention of Cruelty to Animals to be established in Rome, on the grounds that to do so would imply that human beings have duties toward animals.[48] As late as 1951, the American Catholic philosopher V. J. Bourke, in a textbook on ethics, justifies eating meat by restating Aquinas, who was already restating Aristotle:

> In the order of nature, the imperfect is for the sake of the perfect, the irrational is to serve the rational. Man, as a rational animal, is permitted to use things below him in this order of nature for his proper needs. He needs to eat plants and animals to maintain his life and strength. To eat plants and animals, they must be killed. So killing is not, of itself, an immoral or unjust act.[49]

We saw earlier that Aquinas had sufficient reason to know that eating animals is not necessary for maintaining human life. Bourke sticks so closely to Aquinas that he repeats the same claim, when he would only have needed to look up a standard work on nutrition to discover that it was false.

Still, by the end of the twentieth century, there were signs that the environmental movement was beginning to affect the Church's interpretation of the supposed grant to humans of dominion over the other animals. Pope John Paul II, in his 1988 encyclical *Sollicitudo rei socialis* (On Social Concerns), urged that human development should include "respect for the beings which constitute the natural world" and added:

> The dominion granted to man by the Creator is not an absolute power, nor can one speak of a freedom to "use

> and misuse," or to dispose of things as one pleases. . . . When it comes to the natural world, we are subject not only to biological laws, but also to moral ones, which cannot be violated with impunity.[50]

A few years later, Cardinal Joseph Ratzinger, when head of the Sacred Congregation for the Doctrine of the Faith, gave an interview to a journalist who asked him about human duties toward animals. Ratzinger gave a surprisingly strong reply, in the course of which he condemned not only the force-feeding of geese to produce pâté de foie gras but also industrialized farming, mentioning "hens so packed together that they just become caricatures of birds." This was, he said, the "degrading of living creatures to a commodity," which seemed to him to "contradict the relationship of mutuality that comes across in the Bible."[51] Ratzinger later became Pope Benedict XVI, a position in which some hoped he would tell Catholics that they should not be eating pâté de foie gras, or the flesh of "living creatures degraded to a commodity;" but he made no further public statements about factory-farmed animals.

Expectations were aroused again in 2013, when Cardinal Bergoglio was elected pope and took the name Francis, in honor of Francis of Assisi, the patron saint of animals. The following year Francis caused a flutter when he said, in the course of remarks about the afterlife, "Holy Scripture teaches us that the fulfillment of this wonderful design also affects everything around us." Did the pope mean, people asked, that animals can go to heaven? Theologians warned that Francis was speaking casually, rather than making a doctrinal statement.[52] That could not be said in 2015, however, when Francis released the encyclical *Laudato si'*. The title is drawn from "The Canticle of the Sun," a song written by

Francis of Assisi, and the encyclical emphasizes the need to care for the natural environment. In that context it includes an explicit disavowal of the Thomistic reading of man's dominion:

> Although it is true that we Christians have at times incorrectly interpreted the Scriptures, nowadays we must forcefully reject the notion that our being created in God's image and given dominion over the earth justifies absolute domination over other creatures.

Laudato si' even repudiates St. Paul's interpretation of the biblical command to rest one's ox on the sabbath, which Paul thought was made "altogether for our sakes." Not so, says Pope Francis: "Rest on the seventh day is meant not only for human beings, but also so that your ox and your donkey may have rest' (Exodus 23:12)." To make sure that the point is not missed, the pope adds: "Clearly, the Bible has no place for a tyrannical anthropocentrism unconcerned for other creatures."

The encyclical calls on us to recognize that "other living beings have a value of their own in God's eyes" and points to the Catechism—a statement of the Church's essential teachings—where it says that we must "respect the particular goodness of every creature, to avoid any disordered use of things." These and other statements in *Laudato si'* signal a historic and much-needed change of direction in Catholic teaching about animals and the environment. Yet although Church leaders are often ready to tell people how to behave in their sexual lives, neither the pope nor any other high-ranking Church official has suggested that to consume the products of factory farms is wrong because these methods of producing animals are disordered uses of animals that fail to respect the particular goodness of those animals.

BEYOND SPECIESISM

CAN CHRISTIANITY REDEEM ITS PAST AND BECOME A NON-speciesist religion? It won't be easy, but there are signs of hope in a group of Christians, both Roman Catholic and Protestant, who are finding sufficient basis in the Christian tradition for a more far-reaching change in our thinking about animals. Andrew Linzey, an English Anglican priest and the founding director of the Oxford Centre for Animal Ethics, led the way with his *Animal Rights: A Christian Perspective*, published in 1976. Charles Camosy, who has taught theological and social ethics at two Jesuit universities, Fordham and Creighton, wrote *For Love of Animals: Christian Ethics, Consistent Action.* As the book's subtitle suggests, Camosy is prepared to tackle, in a thoughtful and honest way, a question to which most Catholic thinkers on this topic have offered manifestly flawed answers: What should a Christian eat? The most comprehensive of these new Christian works about animal ethics is theology professor (and Methodist) David Clough's 2014 book *On Animals*, a thoroughly non-speciesist two-volume work that runs to more than five hundred pages. Clough denies that humans are the center or the purpose of creation and urges that Christians should not work in intensive farms, should oppose the harmful use of animals in research, and have "strong faith-based reasons to adopt a vegan diet."

Meanwhile the secular utilitarian tradition, which for much of the twentieth century seemed to have lost the reforming zeal it had in the early nineteenth century, has gained new and more active adherents in the late twentieth and early twenty-first century. Utilitarianism is consistent with the views defended in this book—which is not a coincidence, given that its author is a utilitarian—but at no point do the arguments I have presented require acceptance of utilitarianism. That speciesism is wrong, and that the practices described in chapters 2 and 3 of this book are morally indefensible, are now accepted by leading representatives of most of the ethical positions held by Western philosophers today, whether they are advocates of animal rights, like Tom Regan; of social contract ethics, like Mark Rowlands; of feminist philosophy, like Carol Adams, Alice Crary and Lori Gruen; Kantians, like Christine Korsgaard; or working within the capabilities approach, like Martha Nussbaum. That philosophers with such diverse approaches to ethics agree that we need a fundamental change in our attitudes toward and our treatment of animals is a strongly positive sign[53]

At least in Western ethical thinking, there are grounds for believing that the dominance of speciesism as an ideology has come to an end. How much impact this shift in ethical thinking will eventually have on the way animals are treated remains to be seen. Although contemporary attitudes toward animals are sufficiently benevolent—on a very selective basis—to allow some improvements in the conditions of animals to be made without challenging basic attitudes toward animals, it is only by making a radical break with more than two thousand years of Western thought about animals that we can build a solid foundation for a world without speciesism. We should not forget, however, that this chapter has focused on Western ideas about animals. Thinkers in China, for example, appear to be less concerned about the ethical issues arising from eating meat, despite the influence there of the Buddhist tradition, which one might expect to lead to greater concern.[54] In the next chapter we will look at some improvements in the treatment of animals that the animal movement has already brought about, and the prospects for further progress.

CHAPTER 6

Speciesism Today . . .

objections to Animal Liberation, and the progress made in overcoming them

GIVE ME THE CHILD UNTIL THEY ARE SEVEN . . .

WE HAVE SEEN HOW, IN VIOLATION OF THE FUNDAMENTAL MORAL principle of equal consideration of similar interests that ought to govern our relations with all sentient beings, humans will inflict suffering on nonhuman animals for trivial purposes. We have also seen how many generations of Western thinkers have sought to defend the right of human beings to do this. I shall use this final chapter to look at some of the ways in which speciesist practices are maintained and promoted today, and at the various arguments and excuses that are still used in defense of our ruthless exploitation of animals. In doing so I hope to answer some of the objections most often made to the case against speciesism. I will then end on an encouraging note by describing some of the progress made for animals since 1975.

Children have a natural love of animals, and parents encourage them to be affectionate toward companion animals such as dogs and cats. They give them cute stuffed toy animals to cuddle. When meat is first introduced into a child's food, some children at first refuse to eat it. They become accustomed to it only after strenuous efforts by their parents, many of whom mistakenly believe that it is necessary for good health. Whatever the child's initial reaction, though, most of us became used to eating animal flesh long before we were capable of understanding that what we were eating was the dead body of an animal. Thus we never made a conscious, informed decision, free from the bias that accompanies any long-established habit and the pressures of social conformity, to eat animal flesh. Jesuit educators supposedly said: "Give me the child until he is seven and I will show you the man." Fortunately today's adolescents have more opportunity to gain different perspectives, so that is no longer true, but even so, after many years of eating meat, it is not easy to think completely differently about the animals one is eating.

When farmed animals are mentioned in picture books, stories, and on television and video programs for children, there is no mention of the real conditions in which they are likely to live. In this respect, nothing has changed since the earlier editions of this book were written. Then, as now, popular books about these animals present children with pictures of hens, turkeys, cows, and pigs, all surrounded by their young, with not a cage, shed, or stall in sight. As I revise this chapter, one of the bestselling books in the children's category is Amy Pixton's *Hello Farm!* Amidst colorful illustrations of happy pigs outside rolling in the mud, and hens pecking at the ground with their chicks beside them, the text asks:

Who rolls in the mud? Playful pigs!
Who greets the day? Crowing roosters!

No wonder that children grow up believing that farms look after their animals well![1]

In the 1970s the feminist movement had some success in fostering the growth of a new children's literature in which brave princesses rescue helpless princes and girls play some of the central, active roles that used to be reserved for boys. To make what we read to our children about animals convey some of today's reality will be more difficult, since cruelty is not an ideal subject for children's stories. Yet it should be possible to avoid the most gruesome details and still give children picture books, stories, and videos that encourage respect for animals as independent beings, and not merely as cute little objects that exist for our amusement and our dinner table; and as children grow older, they can be made aware of the real conditions under which most farmed animals live. The difficulty is that in meat-eating families, parents will be reluctant to let their children learn the truth, for fear that the child's affection for animals may disrupt family meals. Even now, one frequently hears that, upon learning where meat comes from, a friend's child has refused to eat meat. This instinctive rebellion is likely to meet strong resistance, and most children are unable to keep up their refusal in the face of opposition from parents who provide their meals and tell them that they will not grow up big and strong without meat. Here's an instructive example. Lawrence Kohlberg, a Harvard professor of psychology noted for his work on moral development, related in one of his essays how his son, at the age of four, made his first moral commitment: He refused to eat meat because "it's bad to kill animals." It took Kohlberg six months to talk his son out of this position, which, Kohlberg says, was based on a failure to make a proper distinction between justified and unjustified killing and indicates that his son was only at the most primitive stage of moral development.[2]

MAINTAINING IGNORANCE

IT IS AN INDICATION OF THE EXTENT TO WHICH PEOPLE ARE NOW isolated from the animals they eat that children brought up on books picturing idyllic farms could live out their entire lives without ever being forced to revise this benign image. On a drive through the country one now sees many farm buildings and relatively few animals out in the fields, but how many of us can distinguish a storage barn from sheds housing laying hens in cages, or envisage what life is like for the animals inside those sheds? Nor does the mass media educate the public on this topic. It is easy to find popular television programs on animals in the wild, but films of intensive farms are limited to the briefest of glimpses as part of infrequent "specials" on agriculture or food production. The average viewer must know more about the lives of cheetahs and sharks than they know about the lives of chickens or pigs. The result is that most of the "information" about farm animals to be gained from watching television is in the form of paid advertising, which ranges from ridiculous cartoons of pigs who want to be made into sausages and tuna trying to get themselves canned, to images of pigs outdoors in the fields, frolicking in hay. One advertisement for sausages sold by the British supermarket chain Tesco showed two pigs following a farmer down a dirt track between trees. The pig products Tesco was promoting were, in fact, from pigs reared entirely indoors, with thousands of other pigs, and when viewers complained, Tesco had to withdraw the advertisement.[3] Newspapers do little better, giving ample coverage of the birth of a baby gorilla at the zoo but leaving the deliberate mass killing of pigs or chickens from heatstroke largely unreported. Ignorance, therefore, is the speciesist's first line of defense. That ignorance is easily remedied, in this internet age, by anyone with an interest in finding out the truth. But do meat-eaters really want to know how the animals they are eating lived? "Don't tell me, you'll spoil my dinner" is an all-too-common reply to an attempt to tell someone how their dinner was produced.

"HUMANS COME FIRST"

AMONG THE FACTORS THAT MAKE IT DIFFICULT TO AROUSE PUBLIC concern for animals, perhaps the hardest to overcome is the assumption that "humans come first" and that any problem for animals cannot be comparable, as a serious moral or political issue, to problems humans have. A number of things can be said about this assumption. First, it is in itself an indication of speciesism. How can anyone who has not made a thorough study of the topic possibly know that the problem of what we do to animals is less serious than problems of human suffering? One can claim to know this only if one assumes that animals really do not matter, and that no matter the degree of suffering, it is less important than any suffering of humans. But this is, as I have already argued, simply speciesism.

Many problems in the world deserve our time and energy. Famine and poverty, climate change, racism, war and the threat of nuclear annihilation, sexism, unemployment, preservation of our fragile environment—who can say which is the most important? Yet once we put aside speciesist biases we can see that the oppression of nonhumans by humans belongs in the top tier of priorities for change, right alongside these issues. The suffering that we inflict on sentient nonhuman beings can be extreme, and as we have seen, the numbers involved are gigantic, running into the hundreds of billions, and possibly trillions.

In any case, the idea that "humans come first" is more often used as an excuse for not doing anything about either humans or nonhuman animals than as a genuine choice between incompatible alternatives. For the truth is that there is no incompatibility here. Granted, everyone has a limited amount of time and energy, and time devoted to active work for one cause reduces the time available for another cause; but there is nothing to stop those who devote their time and energy to human problems from joining the boycott of the products of agribusiness cruelty. It takes no more time or energy to be a vegetarian or vegan than to eat animal flesh. In fact, as we saw in chapter 4, those who claim to care about the well-being of human beings and the preservation of our climate and our environment should become vegans for those reasons alone. Doing so would reduce greenhouse gas emissions and other forms of pollution, save water and energy, free vast tracts of land for reforestation, and eliminate the most significant incentive for clearing the Amazon and other forests. I do not question the sincerity of vegans who take little interest in animal liberation because they give priority to other causes; but when meat-eaters say that "human problems come first" I cannot help wondering what exactly it is that they are doing for human beings that compels them to continue to support the ruthless exploitation of farmed animals.

A corollary of the idea that "humans come first" is the myth that people in the animal welfare movement care more about animals than they do about human beings. In fact, the leaders of the animal welfare movement typically care far more about human beings than do other humans who care nothing for animals. Indeed, the overlap between leaders of movements against the oppression of racial minorities and women, and leaders of movements against cruelty to animals, is so extensive as to provide an unexpected form of confirmation of the parallels among racism, sexism, and speciesism. Among the handful of founders of the RSPCA, for example, were William Wilberforce and Thomas Fowell Buxton, two of the leaders in the fight against slavery in the British Empire.[4] As for early feminists, Mary Wollstonecraft wrote, in addition to her pioneering *Vindication of the Rights of Woman*, a collection of children's stories titled *Original Stories*, expressly designed to encourage kinder practices toward animals;[5] and a number of the early American feminists, including Lucy Stone, Amelia Bloomer, Susan B. Anthony, and Elizabeth Cady Stanton, were connected with the vegetarian movement. Together with Horace Greeley, the reforming, antislavery editor of the *Tribune*, they would meet to toast "Women's Rights and Vegetarianism."[6]

To the animal welfare movement too must go the credit for starting the fight against cruelty to children. In 1874 Henry Bergh, the founder of the American Society for the Prevention of Cruelty to Animals, was asked by Etta Wheeler, a Methodist missionary, to help a young animal who was being cruelly beaten. The animal was a child named Mary Ellen Wilson. Wheeler approached Bergh because she knew of his success in preventing cruelty to animals, using a New York animal protection statute that he had drafted and bullied the legislature into passing. Legal efforts to prevent parents and custodians abusing children, on the other hand, had been less successful. Bergh successfully prosecuted the child's custodians for cruelty using the same statute. Further cases were then brought, and the New York Society for the Prevention of Cruelty to Children was set up. When the news reached Britain, the RSPCA set up a British counterpart—the National Society for the Prevention of Cruelty to Children.[7] One of the founders of this group was Lord Shaftesbury, a key figure in the passage of the Factory Acts, which put an end to child labor and fourteen-hour workdays. Shaftesbury also campaigned against uncontrolled experimentation on animals. His life, like that of many other humanitarians, refutes the idea that those who care about animals do not care about humans.

OUR MOTIVATED VIEW OF ANIMALS

THE WAY WE THINK ABOUT NONHUMAN ANIMALS HELPS TO BUTtress our speciesist attitudes. As we saw in the previous chapter, Aquinas thought that animals are savage and brutal because they kill us in order to eat us, whereas our killing of them is supposedly just.[8] To say that people are "humane" is to say that they are kind; to say that they are "beastly," "brutal," or simply that they behave "like animals" is to condemn them. As Paley noticed, the truth is just the reverse: Wolves and tigers must kill, or they will starve. Humans kill other animals to please their palates.

While we overlook our own savagery, we exaggerate that of other animals. The notorious wolf for instance, villain of so many folktales, has been shown by the careful investigations of zoologists in the wild to be a highly social animal, a faithful and affectionate spouse—not just for a season, but for life—a devoted parent, and a loyal member of the pack. Wolves almost never kill anything except to eat it. If males should fight among themselves, the fight ends with a gesture of submission in which the loser offers to his conqueror the underside of his neck—the most vulnerable part of his body. With his fangs only an inch away from the jugular vein of his foe, the victor will be content with submission and, unlike a human conqueror, does not kill the vanquished opponent.[9]

In keeping with our picture of the world of animals as a bloody scene of combat, we ignore the extent to which other species exhibit a complex social life, recognizing and relating to other members of their species as individuals. When human beings marry, we attribute their closeness to each other to love, and we feel keenly for a human being who has lost his or her spouse. When other animals pair for life, we say that it is just instinct that makes them do so, and if a hunter or trapper kills or captures an animal for research or for a zoo, we are not disturbed by the thought that the animal might have a partner who will suffer from their loss. In a similar way we know that to part a human mother from her child is tragic for both; but the breeders of companion animals, research animals, and the animals we eat give no thought to the feelings of the nonhuman mothers and children whom they routinely separate as part of their business.[10] Those who treat animals in this manner often brush off criticism by saying that, after all, "We shouldn't anthropomorphize animals." Of course, we cannot assume that nonhuman animals feel as we would in these circumstances, but the dangers of sentimental anthropomorphism are less serious than the opposite danger of the convenient view that animals are mindless machines who feel nothing—when the evidence is clear that animals of many species do feel emotions akin to our love, fear, boredom, loneliness, and grief.

A group of researchers in psychology from Australia and England became interested in exploring what they called "the meat paradox"—the fact that most people like animals and are disturbed by the thought of harming them, yet they eat meat. How, they wondered, do people handle this apparent conflict? The researchers designed an ingenious set of studies in which participants were first asked to say how confident they are that a cow or a sheep has the capacity to experience pleasure, fear, joy, happiness, pain, hunger, pride, and to have desires or wishes. After being distracted with an unrelated task, some participants were told that they would be asked, as part of a study of consumer behavior, to sample some food. Some were told they would be eating lamb, others beef, and still others, apples. After being told this, they were again asked to assess the capacities of a cow or a sheep. Participants who had been told that they would be eating meat were more likely than they had been earlier to deny mental characteristics to the type of animal they expected they would soon be eating, whereas the participants who were expecting to eat apples did not change their assessments of the mental capacities of cows or sheep. The researchers explained this result as an attempt to reduce negative feelings that the participants would otherwise have when they think about the full range of mental capacities that a cow or a sheep has, while expecting soon to be eating meat from that animal. The researchers refer to this as "motivated mind perception." In other words, we tend to deny that the animals we eat have minds because that makes it easier for us to avoid being troubled by eating them.[11] These findings offer new insight into people's ability to object to harming animals while themselves participating in a practice that obviously does harm them. At the same time, the study shows that people have to go through some mental contortions to suppress the fact that they are responsible for harming animals. Perhaps, if people could easily obtain tasty vegan food, without having to resist family or peer pressure, they would see this as a way of reducing the need for these contortions and instead bringing their diet into line with their beliefs.

THEY KILL EACH OTHER, SO WHY SHOULDN'T WE KILL THEM?

AS WE SAW IN THE PREVIOUS CHAPTER, BENJAMIN FRANKLIN APpealed to this argument when the smell of frying fish induced him to break from his vegetarian diet. William Paley rejected it on the grounds that while human beings can live without killing, other animals have no choice but to kill if they are to survive.[12] A few exceptions may be found—animals who could survive without meat but eat it occasionally (chimpanzees, for example)—but they are not the species we commonly find on our dinner tables. More important, though, even if other animals who could live on a vegetarian diet do sometimes kill for food, this would provide no support for the claim that it is morally defensible for us to do the same. It is odd how humans, who normally consider themselves so far above other animals, will support their dietary preferences with an argument that implies that we ought to look to other animals for moral inspiration and guidance. The point, of course, is that nonhuman animals are not capable of considering the alternatives, or of reflecting morally on the rights and wrongs of killing for food; they just do it. We may regret that this is the way the world is, but it makes no sense to hold nonhuman animals morally responsible or culpable for what they do. Every reader of this book, on the other hand, is capable of making a moral choice on this matter. We cannot evade responsibility for our choices by imitating the actions of beings who are incapable of making this kind of choice.

You may object that I have now admitted that there is a significant difference between humans and other animals, and this undermines my case for the equality of all animals. But I have never claimed that there are no significant differences between normal adult humans and other animals. My point is not that animals are capable of acting morally, but that the moral principle of equal consideration of interests applies to them as it applies to humans. That it is often right to include within the sphere of equal consideration beings who are not themselves capable of making moral choices is implied by our treatment of young children and other humans who, for one reason or another, do not have the mental capacity to understand the nature of moral choice. As Bentham might have said, the point is not whether they can choose, but whether they can suffer.

Perhaps the claim is a different one, simply that, as the poet Tennyson put it, nature is "red in tooth and claw" and we are inevitably part of that system in which one preys upon another.[13] But we should be wary of appeals to "nature" in ethical arguments. We must use our own judgment in deciding when to follow nature. Perhaps war is "natural" to human beings—it certainly seems to have been a preoccupation for many societies over a long period of history—but that does not justify starting a war. We have the capacity to reason about what it is best to do, and we should use this capacity. You might say that it is part of our nature to do so.

WILD ANIMAL SUFFERING

THE EXISTENCE OF PREDATORS LIKE LIONS, TIGERS, AND WOLVES does pose a question about including animals within the sphere of our moral concern: If we could eliminate them, thereby reducing the total amount of suffering in the world, should we do it? Initially, this question was asked mostly by those who wanted to show that granting rights to animals leads to absurdity.[14] Today it is being asked by some animal advocates.

One possible answer is that once we cease to claim dominion over the other animals, we should simply leave them alone, and not seek to replace our tyranny with some form of benevolent despotism. This answer is too glib and its implications would often be callous. Should we stop coming to the aid of animals fleeing a wildfire, or trying to help stranded whales return to the ocean? I don't think so. Yes, these actions interfere with nature, but we don't hesitate to do that when humans are at risk from natural events like floods or droughts.

We may be justified in interfering in nature in limited ways, when we can be confident that we will greatly reduce the suffering of wild animals without doing more harm than good in the long run. Eliminating predators, however, would not be the best way to achieve that goal. The American ecologist Aldo Leopold, once a keen hunter of wolves, later came to see that eliminating wolves leads to an increase in the deer population, which in turn causes the deer to overgraze their habitat, resulting in the loss of other species.[15] Eventually, the population of deer will be controlled by their food supply, and when that runs out, they will starve to death—a slower and often more distressing death than being killed by a wolf. Although mentioning wild animal suffering brings to mind images of lions killing zebras, neither Animal Ethics (www.animal-ethics.org) nor Wild Animal Initiative (www.wildanimalinitiative.org), the most prominent organizations primarily concerned with the question of wild animal suffering, supports proposals to eliminate predators. Instead they advocate the development of the new interdisciplinary field of "welfare biology," the aim of which would be to study factors that affect the suffering of animals living in the wild, and assess interventions that could reduce this suffering with no more than minimal risk of negative consequences.[16] Some such interventions would be feasible now. There have already been successful programs to vaccinate wild animals against diseases such as rabies, tuberculosis, and swine flu, putting the vaccine into food that the animals will eat, and then distributing the food, often by dropping it from planes and helicopters. These programs were implemented because the diseases posed a threat to humans or to domesticated animals, but similar techniques could be used to reduce the suffering of the affected animals. Because humans have already transformed urban and suburban areas, interventions to benefit free-living animals there may be more acceptable than programs in wilderness areas. They could advance our understanding of the welfare of free-living animals and how we may be able to assist them.[17]

Other things that we could easily do include protecting birds by encouraging (or, in high-risk areas, requiring) the use of bird-friendly glass in windows, to reduce the danger that birds will fly into windows. Ensuring that domestic cats are not a threat to birds and other small animals would save billions of lives every year. Fireworks cause severe, and sometimes fatal, stress to wild animals and should not be permitted in areas with significant numbers of them. Projects interfering with nature, such as road construction or planned burns to reduce the risk of wildfires, should be required to take into consideration, and minimize, harm to animals. Regulations governing self-driving cars should require them to be programed to avoid hitting wild animals whenever possible, as they are now programed to avoid hitting pedestrians. Perhaps the most important way of reducing the suffering of wild animals is to establish extensive marine sanctuaries that prohibit commercial and recreational fishing, which is a direct, ongoing attack, on a gigantic scale, on wild, free-living animals.[18]

Predation is unlikely to be the greatest cause of wild animal suffering. It may be surpassed by the vast numbers of deaths that occur among species that, in contrast to us and many other birds and mammals, have very many offspring, very few of whom survive to maturity. Frogs, for example may have thousands of offspring, and a single sunfish may lay up to 300 million eggs, of which a few million may be fertilized and hatch.[19] The populations of these animals are not exploding, so we know that, as a long-term average, roughly two of these offspring will themselves survive and reproduce.

Some philosophers believe that all these premature deaths

mean that wild animals, taken as a whole, experience more suffering than happiness.[20] It is hard to know if this pessimistic view is true. If tadpoles hatch in a puddle, swim freely for a few days and then die when the puddle dries up, do their lives have more pain than pleasure? We can ask similar questions of young fish who starve or are eaten by predators, and of insects, but with insects there is the added uncertainty about whether they are even capable of suffering.

Any proposal to reduce the suffering of wild animals is bound to give rise to fears that this goal will be in conflict with the preservation of the natural environment. Yet regarding animals as sentient beings with a stake in the preservation and integrity of the biotic community in which they live may help, rather than hinder, environmental conservation by giving us a different, less human-centered, perspective on what occurs when we try to restore an ecological system. Knowledge of how healthy animals are, and of how they are behaving, can increase our understanding of the factors that cause them to thrive or, conversely, to die out. Moreover, because many people have empathy for wild animals, acknowledging the importance of the well-being of wild animals may lead to more people being involved in protecting nature and animals.[21] In addition, those concerned with reducing the suffering of wild animals will join with conservationists in trying to protect endangered animals like elephants, rhinoceroses, hippopotamuses, tapirs, and great apes. These large animals are herbivores (or in the case of chimpanzees and bonobos, primarily herbivorous). They have few offspring, and they take care of each of them. Because of their size, they are typically not threatened by predators other than humans. Plausibly, they have a positive quality of life. If they should become extinct, it is likely that they would be replaced by smaller animals, more vulnerable to predators, and therefore likely to experience more suffering. Concern for wild animal suffering therefore gives us an additional reason to protect these large herbivores.

BETTER A CHICKEN THAN A JUNGLE FOWL?

IF LIFE FOR AN ANIMAL IN THE WILD IS LIKELY TO BE PAINFUL and short, then isn't it true that, bad as modern farm conditions are, they may still be better than conditions in the wild? At least in a factory farm, most animals are protected from predators and extreme heat and cold. The implication is that we should not object to modern farm conditions if they provide even a modest improvement on conditions in the wild.

It is difficult to compare two sets of conditions as different as those in the wild and those on a factory farm, but if the comparison has to be made, a shorter and riskier life of freedom is likely to be better than one in conditions that prevent animals from walking, running, stretching freely, and being part of a social group suited to the animals' needs. The steady supply of food on a farm is not an unmitigated blessing, since it deprives animals of their most basic natural activity, the search for food. The result for many factory farmed animals is a life of utter boredom, with nothing to do for most of the day except lie in a stall. Nor is it true that all farmed animals get enough food and are protected from extremes of heat and cold. As we have seen, the parents of the chickens and pigs fattened for market are kept permanently hungry; cows in feedlots must stand without shade in summer, and some of them die from heatstroke. Others die of cold during winter blizzards. As for protection from extreme heat, as we saw in chapter 3, on the contrary, factory-farmed pigs and chickens may be deliberately overheated to death.

In any case, the comparison between factory-farm conditions

and natural conditions is irrelevant to the justifiability of factory farms, since this is not the choice that we face. Animals in factory farms today were bred by humans to be raised in these farms and sold for food. If we stop eating factory-farmed animal products, that will make raising animals on factory farms less profitable. Farmers will turn to other types of farming, and the giant corporations will invest their capital elsewhere. The result will be that fewer animals will be bred. No animals will be "returned" to the wild. Eventually, perhaps (and now I am allowing my optimism free rein), the only herds of cattle and pigs to be found will be on large reservations, rather like our wildlife refuges. The choice, therefore, is not between life on a factory farm and life in the wild, but between life on a factory farm and not being born at all. To bring them into existence for the kind of life they lead is no benefit to them but rather a great harm.

THE INCONSISTENCY OBJECTION

SPECIESISM IS SO PERVASIVE AND WIDESPREAD AN ATTITUDE THAT those who attack one or two of its manifestations—like hunting or cruel experimentation, or bullfighting—often participate in other speciesist practices themselves. This allows those attacked to accuse their opponents of inconsistency. "You say we are cruel because we shoot deer," the hunters say, "but you eat meat. What is the difference, except that you pay someone else to do the killing for you?" "You object to killing animals to clothe ourselves in their skins," say the furriers, "but you are wearing leather shoes." Experimenters plausibly ask why, if people accept the killing of animals to please their palates, they should object to the killing of animals to advance knowledge; and if the objection is just to suffering, they can point out that animals killed for food do not live without suffering either. Even the bullfighting enthusiast can argue that the death of the bull in the ring gives pleasure to thousands of spectators, while the death of the cow in a slaughterhouse gives pleasure only to the few who eat some part of it; and while in the end the bull may suffer more acute pain than the cow, for most of his life it is the bull who is better treated.

As the English novelist and animal advocate Brigid Brophy once put it, it remains true that it is cruel to break people's legs, even if the statement is made by someone in the habit of breaking people's arms.[22] Yet people whose conduct is inconsistent with their professed beliefs will find it difficult to persuade others that their beliefs are right; and they will find it even more difficult to persuade others to act on those beliefs. Of course, it is always possible to find some reason for distinguishing between, say, wearing furs and wearing leather: Many fur-bearing animals die only after hours or even days spent with a leg caught in a trap, while the animals from whose skins leather is made are spared this agony.[23] There is a tendency, however, for these fine distinctions to blunt the force of the original criticism; and in some cases I do not think distinctions can validly be drawn at all. Why, for instance, is the hunter who shoots a deer for venison subject to more criticism than the person who buys a ham at the supermarket? If the hunter can shoot well enough to kill the deer quickly, it will be the intensively reared pig who has suffered more.

The first chapter of this book sets out a clear ethical principle by which we can determine which of our practices affecting nonhuman animals are justifiable and which are not. By giving equal weight to the similar interests of all sentient beings, irrespective of their species, we can make our actions fully consistent. Thus we can deny to those who ignore the interests of animals the opportunity to charge us with inconsistency. For all practical purposes as far as urban and suburban inhabitants of the industrialized nations are concerned, following the principle of equal consideration of interests requires us, at a minimum, to avoid factory-farmed animal products, and many of us will want to take this further and avoid eating animal products altogether. We should also, to be consistent, stop using other animal products for which animals have been killed or made to suffer. We should not wear furs. We should avoid leather products too, since the sale of hides for leather plays a significant role in the profitability of the meat industry.

For the pioneer vegetarians of the nineteenth century, giving up leather meant a real sacrifice, since shoes and boots made of other materials were scarce and often inadequate. Lewis Gompertz, the second secretary of the RSPCA and a strict vegetarian who refused to ride in horse-drawn vehicles, suggested that animals should be reared in pastures and allowed to grow old and die a natural death, after which their skins would be used for leather.[24] The idea is a tribute to Gompertz's humanity rather than his economics, but today the economics are on the other foot. Shoes and boots made of canvas, rubber, and synthetics are now available in many stores, both in low-priced stores and in the high-end designer ranges. Belts, bags, and other goods once made of leather are now easily found in other materials too.

Other problems that used to daunt the most advanced opponents of the exploitation of animals have also disappeared. Candles, once made only of tallow, are no longer indispensable and can, for those who still want them, be obtained in nonanimal materials. Soaps made from vegetable oils rather than animal fats are on sale everywhere. We can do without wool, and although sheep generally roam freely, there is a strong case for avoiding wool in view of the many cruelties to which these gentle animals are subjected.[25] Cruelty-free cosmetics and perfumes that do not contain animal products and have not been tested on animals either are now widely available.

Although I mention these alternatives to animal products to show that it is not difficult to refuse to participate in the major kinds of exploitation of animals, I do not believe that consistency requires standards of absolute purity in all that one consumes or wears. The point of avoiding animal products is not to keep yourself untouched by evil but to reduce your economic support for the exploitation of animals and to persuade others to do the same. So it is not a sin to continue to wear leather shoes you bought before you began to think about these issues. You will not reduce the profitability of killing animals by throwing them out. Wear them out and then buy nonleather ones. With diet, also, depending on the circumstances and your relationship with your host, there may be more at stake than worrying about such details as whether the cake you are offered at a party contains eggs. We are more likely to persuade others to share our attitude if we temper our ideals with common sense than if we strive for the kind of purity that is more appropriate to a religious dietary law than to an ethical and political movement.

Usually it is not too difficult to be consistent in one's attitudes toward animals. We need to acknowledge, though, that some clashes of interests between ourselves and animals are unavoidable. We need to grow grain, legumes and vegetables, and fruits to feed ourselves; but these crops may be threatened by rabbits, mice, or other "pests." What can be done about that if we act in accordance with the principle of equal consideration of interests?

First let us note what is done about this situation now. The farmer will seek to kill off the "pests" by the cheapest method available, commonly a poison. How long it takes the animals to die after eating the poison will depend on the type of poison used and the species of the animal, but in some cases it will be a slow, painful death. The very word "pest" seems to exclude any concern for the animals themselves. But the classification "pest" is our own, and a rabbit seen as a pest is as capable of suffering, and as deserving of consideration, as a rabbit who is a beloved companion animal. The problem is how to defend our own essential food supplies while respecting the interests of these animals to the greatest extent possible. It should not be beyond our technological abilities to find a solution to this problem that, if not totally satisfactory to all concerned, at least causes far less suffering than the present "solution." The use of baits that kill animals quickly and humanely would be an obvious improvement. Better still would be a substance designed to cause sterility, instead of a lingering death. Perhaps, in time, people will come to see that even animals who are in some sense "threatening" our welfare, whether in the fields where we grow our food or in our own homes where we store it, do not deserve the cruel deaths we inflict upon them; and so we may eventually develop and use humane methods of limiting the numbers of those animals whose interests are genuinely incompatible with our own.

WHAT ABOUT PLANTS?

I HAVE ARGUED THAT MUCH OF THE WAY IN WHICH WE TREAT animals—including producing them for food—is wrong because we inflict pain and suffering on them. At this point someone is bound to ask: "How do we know that cutting a lettuce does not inflict pain and suffering too?" Those who ask this may be genuinely uncertain about the consciousness of plants, or they may be motivated by the thought that if it could be shown that plants can suffer, then we may as well continue to eat animals because the only alternative would be starvation. If it is impossible to live without violating the principle of equal consideration, they appear to believe, we may as well go on as we have always done, eating plants and animals.

In earlier editions of this book I summarily dismissed the suggestion that plants may be able to feel pain, saying that nothing resembling a central nervous system has been found in plants and there is no observable plant behavior suggesting that they can feel pain. The first part of this statement remains true, at least in the strict sense of a nervous system. Plants do not have a system of nerves leading to a central brain; but plants do engage in electrical and chemical signaling of various kinds. The response of the Venus flytrap to an insect landing in its trap is well known and studied, but there are many other ways in which plants respond to stimuli—for example, to insects feeding on them. This behavior can change over time in ways that can be seen as showing that the plant has learned what does or does not benefit it. Plants also communicate with other plants. We can regard these responses as intelligent. This advance in our understanding of plants shows that they are not the passive objects that we may imagine they are.[26] But this does not show that they are conscious, or capable of feeling pleasure or pain. Self-driving cars communicate by sending electrical signals, are intelligent, and can learn from their mistakes, but they are not conscious. It's true that plants are natural, evolved living beings, whereas self-driving cars are inanimate objects made by humans. That, however, is not sufficient reason for concluding that plants are conscious. If we can design objects that respond intelligently to their circumstances, but without being conscious, then hundreds of millions of years of evolution could produce a similar outcome.[27]

I am therefore now more open to the possibility that plants can feel pain than I was in the past, but I still think it is very unlikely—and significantly less likely than that oysters can feel pain. But assume that, improbable as it seems, researchers do turn up evidence suggesting that plants feel pain. It would still not follow that we may as well eat animals. If we must inflict pain or starve, we would then have to choose the lesser evil, and minimize the pain we inflict. Presumably it would still be true that plants, lacking brains and centralized nervous systems, suffer less than animals, especially vertebrates, and therefore it would still be better to eat plants than to eat animals. Indeed this conclusion would follow even if plants were as sensitive as animals, since as we saw in chapter 4, those who eat meat are responsible for the indirect destruction of vastly more plants than vegans are. Those who raise this objection but fail to follow out its implications in regard to this well-known fact are really just looking for an excuse to go on eating meat.

SPECIESIST PHILOSOPHY

UP TO THIS POINT WE HAVE BEEN EXAMINING, IN THIS CHAPTER, attitudes that are shared by many people in Western societies, and the strategies and arguments that are commonly used to defend these attitudes. We have seen that from a logical point of view these strategies and arguments are very weak. They are rationalizations and excuses rather than arguments. It might be thought, however, that their weakness is due to some lack of expertise in discussing ethical questions. For that reason, in the first edition of this book I examined what some of the leading philosophers of the 1960s and early 1970s had said about the moral status of nonhuman animals. The results were not to the credit of philosophy.

Philosophy ought to question the basic assumptions of the age. Thinking through, critically and carefully, what most of us take for granted is, I believe, the chief task of philosophy and the task that makes philosophy a worthwhile activity. Regrettably, philosophy does not always live up to its historic role. Aristotle's defense of slavery will always stand as a reminder that philosophers are subject to all the preconceptions of the society to which they belong. Sometimes they succeed in breaking free of the prevailing ideology; more often they become its most sophisticated defenders.

The philosophers who tackled problems that touched upon the issue of the moral status of animals around the time of the initial publication of this book did not challenge anyone's preconceptions about our relations with other species. They made the same unquestioned assumptions as those who knew nothing of philosophy, and what they said tended to confirm readers in their comfortable speciesist habits.

In twentieth-century philosophy, equality and human rights were central topics in ethics and political philosophy, as they are today. No major philosopher directly addressed questions about equality or rights for animals, but in discussing the issue of human equality, it was difficult to avoid entirely the question of the status of animals. For philosophers of the 1950s and 1960s, the problem was to interpret the idea that all human beings are equal in a manner that does not make it plainly false. In most ways, human beings are not equal; and if we seek some characteristic that all of them possess, then this must be a common denominator, pitched so low that no humans lack it. The catch is that any such common denominator that is possessed by *all* humans will not be possessed *only* by humans. For example, only human beings are capable of solving complex mathematical problems, but not all humans can do this; while all humans, but not only humans, are capable of feeling pain. So it turns out that in the only sense in which we can truly say, as an assertion of fact, that all humans are equal, at least some members of other species are also "equal"—equal, that is, to humans. This is, as I have argued, the correct understanding of equality, and it leads to the principle of equal consideration of interests, which includes animals. But philosophers prior to the 1970s could not conceive that this outcome could be correct. Rather than accept it, they sought to reconcile their belief in human equality with their belief that equality does not include animals. In pursuit of this goal they accepted arguments that are either devious or myopic.

One philosopher prominent in philosophical discussions of equality at the time was Richard Wasserstrom, then professor of philosophy and law at UCLA. In an article from 1970, "Rights, Human Rights and Racial Discrimination," Wasserstrom defined "human rights" as those rights that human beings have and nonhumans do not have. He then argued that there are human rights to well-being and to freedom. In defending the idea of a human right to well-being, Wasserstrom said that to deny someone relief from acute physical pain makes it impossible for that person to live a full or satisfying life. He then went on: "In a real sense, the enjoyment of these goods differentiates human from nonhuman entities."[28] The problem is that when we look back to find to what the expression "these goods" refers, the only example given is relief from acute physical pain—something that nonhumans may appreciate as well as humans. So if human beings have a right to relief from acute physical pain, it would not be a specifically human right, in the sense Wasserstrom had defined. Animals would have it too.

Otherwise excellent philosophers were faced with a situation in which they saw a need to provide a basis for the moral gulf that is thought to separate human beings and animals but were unable to find any concrete difference that would do this without implying that some humans fall outside the sphere of equality. They resorted to high-sounding phrases like "the intrinsic dignity of the human individual" or "the intrinsic worth of all men" (sexism was as little questioned as speciesism), as if all humans had some unspecified worth that other beings lack. Or they would say that human beings, and only human beings, are "ends in themselves" while "everything other than a person can only have value for a person."[29]

As we saw in the preceding chapter, the idea of a distinctive human dignity and worth has a long history. Twentieth-century philosophers, until the 1970s, had cast off the original metaphysical and religious shackles of this idea, and freely invoked it without feeling any need to justify the idea at all. Why should we not attribute "intrinsic dignity" or "intrinsic worth" to ourselves? Why should we not say that we are the only beings in the universe who have intrinsic value? Our fellow humans are unlikely to reject the accolades we so generously bestow upon them, and those to whom we deny the honor are unable to object. Indeed, when we think only of human beings, it can be very liberal, very progressive, to talk of the dignity of all of them. In so doing we implicitly condemn slavery, racism, and other violations of human rights. We admit that we ourselves are in some fundamental sense on a par with the poorest, most ignorant members of our own species. It is only when we think of human beings as no more than a small subgroup of all the beings who inhabit our planet that we may realize that in elevating our own species we are at the same time lowering the relative status of all other species.

The truth is that the appeal to the intrinsic dignity of human beings appears to solve the egalitarian philosopher's problems only as long as it goes unchallenged. Once we ask why it should be that all human beings—ranging from anencephalic infants to criminal psychopaths and mass murderers like Hitler and Stalin—have some kind of dignity or worth that no chimpanzee, dog, elephant, horse, or whale can ever achieve, we see that this question is as difficult to answer as our original request for some relevant fact that justifies the superior moral status of humans. In fact, these two questions are really one: Talk of intrinsic dignity or moral worth does not help, because any satisfactory defense of the claim that all and only human beings have intrinsic dignity or worth would need to base that claim on a relevant capacity or characteristic that all humans—and no nonhuman animals—possess. To introduce ideas of dignity and worth as a substitute for other reasons for distinguishing humans and animals is not good enough. Fine phrases such as these are the last resource of those who have run out of arguments.

Could there be some morally relevant characteristic that distinguishes all human beings from all members of other species? The problem, remember, is that there are some human beings whose cognitive capacities—including their level of awareness, self-consciousness, intelligence, and sentience—is below that of many nonhuman animals. Infant humans are in this category, but their potential might be considered morally relevant. There are also, however, humans born with genetic conditions, a chromosomal abnormality, or irreparable brain damage who are profoundly and permanently intellectually disabled. Philosophers who set out to find a characteristic that would distinguish human beings from other animals rarely take the course of abandoning these humans by placing them in the same moral category as nonhuman animals. The reason for not doing so is obvious: To take this line without rethinking our attitudes to other animals suggests that we could perform painful experiments on these profoundly intellectually disabled humans for the same reasons that we now perform them on nonhuman animals. Even more grotesquely, it would imply that it would not be wrong to fatten them, kill them, and eat them, should we wish to do that.

For many years, for philosophers discussing the problem of equality, the easiest way out of this difficulty was to ignore it. The

Harvard philosopher John Rawls, in the most celebrated work of late twentieth-century political philosophy, *A Theory of Justice*, came up against this problem when trying to explain why we owe justice to human beings but not to other animals. He brushed it aside with the remark, "I cannot examine this problem here, but I assume that the account of equality would not be materially affected."[30] That is a questionable assumption, when the problem appears to have only two possible solutions, both of them radically different from the status quo: Either we may treat people who are profoundly and permanently disabled intellectually as we now treat animals, or we owe justice to animals.

What else could philosophers do? In 2006, Bernard Williams, nearing the end of a distinguished career that included appointments to the most prestigious chairs of moral philosophy at both Cambridge and Oxford universities, wrote an essay called "The Human Prejudice"—a title that accurately describes the position Williams defends. Nothing, he said, can be of absolute importance, independently of the point of view of some particular being or group of beings. Ethics is, in other words, relative to the group with which one identifies, and if that group is one's species, then there is nothing wrong with having a prejudice in favor of one's own species. Williams defends this position by imagining that aliens arrive on earth. They turn out to be "benevolent and fair-minded and far-sighted," but their knowledge of us and our ways leads them to seek to remove us, for the benefit of others in this part of the universe. At this point, Williams claimed, the project of trying to sort things out in ways that all rational beings can understand breaks down and "there seems to be only one question left to ask: Which side are you on?"[31]

"Which side are you on?" appeals to some of our worst instincts. Wherever there is racial or ethnic violence, and a member of the dominant group that is inflicting the violence tries to dissuade their fellow Whites, Nazis, or Hutus from attacking Blacks, Jews, or Tutsis, that question will be asked. If it is, "I am one of you, and therefore I am on your side" is precisely the wrong way of answering it. That answer abandons the attempt to solve problems in the light of justice and reason, leaving the resolution of the disagreement to force.

As I mentioned in chapter 1, the Australian philosopher Stanley Benn attempted to save the conventional view of the superior moral status of all humans, even those with cognitive capacities below those of nonhuman animals, by arguing that we should treat beings according to what is "normal for the species" rather than according to their actual characteristics.[32] But why should we do that? In many societies, when couples have children, it is common for women to stay at home looking after the children while men go out to work. Assume that in some society, this occurs even when men and women receive equal pay and in the absence of cultural or social pressure. Nevertheless, in this society, some women prefer working to looking after children, and are better suited to going out to work than the fathers of their children. Would anyone claim that these women should be treated in accordance with what is "normal for the gender"—and they should therefore be required to stay at home with the children while the children's father goes out to work?

I turn now to a more recent and more sophisticated defense of giving a higher moral status to humans. In *How to Count Animals*, Shelly Kagan, a philosopher at Yale University, set out to oppose the view that I have taken in this book—the view that, as he put it, "otherwise similar harms or benefits for people and animals count equally from the moral point of view." Instead, Kagan argued for a hierarchical view: that animals count, but they count for less than people do—and not because humans are capable of being harmed more, or benefiting more, than animals, but because humans have a higher moral status.[33] Kagan's discussion of why humans may have a higher moral status traced a familiar path. He started with the assumption that beings who are persons have a higher moral status, where the term "person" refers to a being who is rational and self-conscious, aware of having a future, and has preferences for that future. Kagan then acknowledges that although infants are not persons, we may nevertheless ascribe higher moral status to infants because of their potential to be a person. That leaves us with humans who lack that potential. Kagan takes up this challenge by asking us to imagine a twenty-year-old who suffered irreparable brain damage as an infant, as a result of which she remains at the cognitive level of a four-month-old, and does not have the potential to become a person. Nevertheless, Kagan says, she may have a special moral status because "she *could have been* a person (now), had the accident not occurred when she was a baby." Kagan calls this "modal personhood" ("modal" here refers to the grammatical mood of verbs indicating possibility, necessity, etc.). The fact that the irreparably brain-damaged human would, but for the accident, now be a person is, Kagan claims, something that can give her a higher moral status than an animal at the same cognitive level because the animal could not have been a person.

Does this mere counterfactual possibility, which we know is not and never will be realized, raise a being's moral status above that which the being has because of its actual characteristics? Kagan doesn't say much in support of the claim that it does, and at the end of his book he admits as much, writing: "All I have really done is to offer some rough, preliminary suggestions about what a suitable account might conceivably look like. The devil will be in the details, and the details are yet to come." In the absence of a more detailed account, I see no reason to accept Kagan's suggestion. One reason against accepting it is that modal personhood draws the line between higher and lower moral status in a very strange place. Imagine two twenty-year-old humans who both have the cognitive capacities of a four-month-old and will never surpass that level. Ann is the one Kagan describes who suffered irreparable brain damage as an infant. The other, Bella, has trisomy-13, which means that she has an extra chromosome 13. Long-term survival with trisomy-13 results in severe and irreparable intellectual disability. As Kagan uses the term "person," Ann could have been a person, but Bella could not have been one because the extra chromosome is part of her genome. If a different egg or sperm had been involved in the conception that took place in her mother's body at that time, a different human being would have been born. If Ann and Bella have identical cognitive capacities and their capacities for relating to others, and generally for suffering or enjoying life, are indistinguishable, we should reject the suggestion that Ann has a higher moral status than Bella. If modal personhood is not a reason for according Ann a higher moral status than Bella, however, it also isn't a reason for thinking that Ann's joys and miseries count more, morally, than the similar joys and miseries of an animal with capacities for suffering or enjoying life that are indistinguishable from Ann's.

The discussion of these questions will continue, as it should, and I can't exclude the possibility that one day someone will offer a convincing case for the view that all humans count for more, morally, than any nonhuman animals. But that doesn't seem likely, in view of the already mentioned difficulties. In the meantime, we should continue to hold that "pain is pain," and from the moral point of view, similar pains should have similar weight, whether they are the pains of humans or animals.

PHILOSOPHY REDUX

KAGAN BEGINS *HOW TO COUNT ANIMALS, MORE OR LESS,* BY NOTing that animal ethics, which was almost completely neglected just fifty years ago, is now a well-established subdiscipline within moral philosophy, and one in which many theorists are drawn to the view (which he opposes) that speciesism is wrong and we should give equal consideration to the similar interests of all sentient beings. That is true, and a sign that philosophy has thrown off its ideological blinkers and returned to its Socratic role of questioning accepted beliefs. The rise of the Animal Liberation movement may be unique among modern social causes in the extent to which it has been linked with the development of the issue as a topic of discussion in the circles of academic philosophy. Many of today's university courses in ethics challenge their students to rethink their attitudes on a range of ethical issues, and the moral status of nonhuman animals is prominent among them. Articles on how we ought to treat animals are included in virtually all the standard collections of readings used in applied ethics courses. At least in Western philosophy, it is the complacent, unargued assumptions that animals do not count, or count less than humans, that have become scarce.

The growing library of books and articles on the ethics of our treatment of animals is only part of the story; in philosophy departments all over the world, and in at least the thirty languages into which this book has now been translated, philosophers are teaching their students about the moral status of animals. Moreover, there is now strong evidence that teaching students about the ethics of eating meat does lead to them eating less meat.[34] Of course, philosophers are not unanimous in their opposition to speciesism or their support of vegetarian or vegan eating—

when were they ever unanimous about anything?—but a study of university professors in Germany, Austria, and Switzerland found that 67 percent of ethicists and 63 percent of philosophers not specializing in ethics considered eating meat from mammals morally bad, which was significantly more than the 39 percent of professors from other disciplines who took this view.[35]

The core of this book is the claim that to discriminate against beings solely on account of their species is a form of prejudice, immoral and indefensible in the same way that discrimination on the basis of race or sex is immoral and indefensible. I have not been content to put forward this claim as a bare assertion, or as a statement of my own personal view, which others may or may not choose to accept. I have argued for it, appealing to reason rather than to emotion or sentiment. I have chosen this path, not because I am unaware of the importance of empathy and other compassionate feelings toward other creatures, but because reason is more universal and more compelling in its appeal. Greatly as I admire those who have eliminated speciesism from their lives purely because their concern for others reaches out to all sentient creatures, I do not think that an appeal to sympathy and compassion alone will convince most people of the wrongness of speciesism. Even where other human beings are concerned, people are surprisingly adept at limiting their sympathies to those of their own nation or race. Almost everyone, however, is at least nominally prepared to listen to reason.

So throughout this book I have relied on rational argument. Unless you can refute the central argument of this book, you should now recognize that speciesism is wrong, and this means that, if you take morality seriously, you should try to eliminate speciesist practices from your own life and oppose them elsewhere. Otherwise no basis remains from which you can, without hypocrisy, criticize racism or sexism.

I have generally avoided arguing that we ought to be kind to animals because cruelty to animals leads to cruelty to human beings. Perhaps it is true that kindness to human beings and to other animals often go together; but to say, as Aquinas and Kant did, that this is the real reason we ought to be kind to animals is a thoroughly speciesist position. We ought to consider the interests of animals because they have interests, and it is unjustifiable to exclude them from the sphere of moral concern. To make this consideration depend on beneficial consequences for human beings is to accept the implication that the interests of animals do not warrant consideration for their own sakes.

MAKING PROGRESS

I BELIEVE THAT THE CASE FOR ANIMAL LIBERATION IS LOGICALLY cogent and cannot be refuted; but the task of overthrowing speciesism in practice is a formidable one. We have seen that speciesism has historical roots that go deep into the consciousness of Western society. We have seen that the elimination of speciesist practices would threaten the vested interests of giant agribusiness corporations with the capacity to bombard the public with advertisements denying allegations of cruelty. Moreover the public has—or thinks it has—an interest in the continuance of the speciesist practice of raising and killing animals for food, and this makes people ready to accept assurances about how the producers care for their animals. As we have seen, people are also ready to accept fallacious forms of reasoning, of the type we have examined in this chapter, which they would never entertain for a moment were it not for the fact that these fallacies appear to justify their preferred diet.

Against these ancient prejudices, powerful vested interests, and ingrained habits, do animal advocates have any chance at all? Do they have anything in their favor, other than reason, morality, and a limited amount of empathy for some animals? We might chart the development of the animal movement by the stages sometimes attributed to Gandhi: "First they ignore you. Then they laugh at you. Then they attack you. Then you win."[36] The early pioneers of the animal movement were largely ignored. When *Animal Liberation* first appeared, many ridiculed it. At the start of the 1980s, the animal movement was still widely seen as crackpot, and the membership of groups with a genuinely anti-speciesist attitude was tiny. By the end of that decade, though, People for the Ethical Treatment of Animals had 250,000 supporters. Henry Spira's coalitions to abolish cruel and unnecessary tests on animals brought together animal rights and animal welfare groups with a combined membership in the millions.[37] At that point the animal movement was obviously a serious force, and it was widely attacked. Today it has still not passed the stage of being attacked, but it also has wide support and has made some significant gains.

Some of these gains for animals have been in areas that can easily be seen to be dispensable, like fur, which nowadays exists only for vanity. The fur industry keeps about 100 million active, curious animals like mink and foxes trapped in small barren wire cages making endless stereotypical movements to relieve their stress. Fur farming has been banned in many countries, including Belgium, Italy, the Netherlands and the United Kingdom, and is restricted in ways that make it economically unviable in Germany. To have an impact on the fur industry in China, which now produces half the world's fur, however, bans on production are not enough.[38] Israel has gone further, banning the sale of fur, and so has the state of California. Many designers, or their brands, including Stella McCartney, Gucci, Versace, Coach,

Chanel, Prada, Burberry, Michael Kors, and Giorgio Armani, have stopped using fur.[39]

Pâté de foie gras is an expensive treat for gourmets who ignore cruelty in order to indulge their palates. As the name indicates, this spread is made from a "fatty liver" but not one that is naturally fatty. The geese or ducks from whom the liver came were force-fed though funnels pushed down their throats until their livers were so engorged that it became difficult for them to stand. The production of foie gras is now prohibited in several European countries, including the United Kingdom; and in the United States, its sale is banned in California. Causing animals to suffer in order to market new cosmetic products is another indefensible area of animal suffering. The European Union has not only prohibited the testing of cosmetics on animals but has also stopped the import of cosmetics, as well as ingredients used only in cosmetics, that have been tested on animals.

Glue traps inflict slow, agonizing, panicked deaths on mice, squirrels, birds, and other small animals who become stuck in the glue and are unable to escape. They have been banned in New Zealand, Ireland, Iceland, some Australian states, three Indian states, and in England, with Scotland and Wales expected to soon ban them as well. Hundreds of corporations and other entities have also stopped using them.[40]

New Zealand banned the export of live animals for slaughter in 2008 but continued to permit the export of dairy cows, sheep, and goats, many of whom suffered severely on long sea voyages across the equator. After continued campaigning by local animal advocates, in 2021 the minister for primary industry announced that the trade would end in 2023. Once the animals left the country, he said, the government was unable to ensure their well-being, and hence live exports posed "an unacceptable risk to

New Zealand's reputation." If only all countries showed as much concern about their reputation for protecting animal welfare!

The Great Ape Project, which I co-founded with the Italian scholar and animal activist Paola Cavalieri in 1993, seeks to narrow the gulf that exists in people's minds between humans and all other animals.[41] In pursuing this aim, we called for our closest relatives—chimpanzees, bonobos, gorillas, and orangutans—to be recognized as self-aware beings with deep social relationships and rich emotional lives, who are therefore entitled to basic rights to life, liberty, and freedom from torture. The latter refers particularly to their use in invasive experiments, which in the 1990s were still commonly carried out on chimpanzees in the United States and some other countries. Since then, harmful experiments on great apes have been prohibited or otherwise ended in the United Kingdom, New Zealand, Japan, and throughout the European Union. In the United States, they ended in 2015, and great apes in scientific facilities have mostly been sent to sanctuaries. In addition, there have been attempts to use legal action, including the writ of habeas corpus, to get great apes out of confinement. In 2016 one of these attempts met with success when a court in Argentina accepted that habeas corpus can apply to a nonhuman animal and freed Cecilia, a chimpanzee.[42]

When writing about factory farming in the first edition, I focused on the three most restrictive forms of confinement then in use: individual stalls for veal calves, barren battery cages for laying hens, and individual stalls for pregnant sows. Animal advocates in the European Union, then including the United Kingdom, succeeded in getting all three of them prohibited, although sow stalls are still allowed for the first four weeks of the sow's pregnancy. The bans were phased in slowly and took effect in 2007, 2012, and 2013, respectively. They allowed more space to hundreds of millions of animals, including room for the calves and sows to turn around; hens are allowed space in which to stretch their wings and have access to a sheltered nest box in which to lay their eggs. These were all important, but incomplete, victories, because the animals could all still be kept in cages, as long as they were allowed relatively free movement within the confined space. In 2021, however, after a campaign involving 170 animal welfare organizations, 140 scientists and professionals in relevant fields, and a petition signed by 1.4 million European citizens, the European Parliament voted overwhelmingly in favor of a ban on all cages for farmed animals. The European Commission agreed to phase out cages for farmed animals across the entire European Union by 2027, a move that will affect more than 300 million animals, including not only calves, pigs, and laying hens, but also rabbits, ducks, geese, quail, and other farmed animals. To prevent the replacement of European products by animal products from countries with lower standards, the commission is planning to introduce equivalent standards for imported animal products.[43]

In Europe, concern about the treatment for animals has led to the founding of political parties specifically campaigning for animals. The most successful so far has been the Party for the Animals, in the Netherlands, which has held seats in the national parliament since 2006. In the general election of 2021, the Party for the Animals received 3.8 percent of the vote, entitling it to 6 of the 150 seats in the lower house of the Dutch parliament and 3 in the senate. Animal advocacy parties based in Germany, Portugal, and Spain have won seats either in their national parliaments or in the European parliament. In Australia, the Animal Justice Party has won seats in the upper houses of the two most populous states, New South Wales and Victoria. Such parties can have influence by speaking up for animals and asking questions in parliament. There is always the chance that in a close election they may hold the balance of power, in which case they could have a more direct influence on legislation and government policy. On the other hand, in countries with "first past the post" voting systems, political parties for animals are unlikely to get their candidates into parliament and they risk drawing votes away from, and so contributing to the defeat of, the major party that has more pro-animal policies. That is the situation in the United States—where Ralph Nader's candidacy for the Green Party in the 2000 presidential election enabled George W. Bush to defeat Al Gore—as well as in the United Kingdom and, for the lower house, Canada and India.

In the United States, agribusiness and farming lobbies have thwarted all attempts to pass federal legislation protecting farm-animal welfare, but there have been important victories for animals at the state level, mostly in states where citizens have the right to initiate referendums. Altogether nine states have bans on either veal stalls, battery cages, or sow stalls. The most impressive wins in referendums have come in California, where 63 percent of voters supported a measure prohibiting confining or tethering sows, veal calves, and laying hens in ways that prevent them from lying down, standing up, turning around, or fully extending their limbs; and in Massachusetts a similar measure was favored by 78 percent of voters. Other states with bans are Oregon, Washington, Colorado, Maine, Utah, Rhode Island, and Nevada.[44]

In the absence of U.S. federal legislation to protect animals being reared for food, animal advocates have achieved important gains by putting pressure on corporations to take into account how animals were treated when they buy meat, eggs, and dairy products. By 2016, all of the top twenty-five supermarket chains in the country had pledged to sell only cage-free eggs by 2026 (although the commitment did not extend to all products containing eggs).[45] In 2016 McDonald's agreed to go cage-free within ten years. (The corporation said that it couldn't make the change earlier, because the 2 billion cage-free eggs it would need every year just weren't available.)[46] Five years later, after a protracted worldwide campaign in which animal organizations from many different countries took part, Yum! Brands—the owner of KFC, Pizza Hut, and Taco Bell, among other brands, and the world's largest fast-food company—pledged that by 2030 it would eliminate eggs from caged hens across all its 50,000 locations in 150 countries.[47] In the U.S. the joint impact of the changes in state legislation and the campaigns to persuade corporations to change their purchasing policies has increased the proportion of hens not in cages from only 3 percent in 2005 to 35 percent in 2022, and that should continue to increase as more corporations meet their commitments.[48] (On the other hand, remember that cage-free hens may be living with thousands of other hens in the ammonia-filled air of a crowded shed, never able to go outside in fresh air and sunshine. It's better than cages, but still a long way from free range or pasture-raised hens.)

AN ALTERNATIVE STRATEGY

DESPITE ALL THESE SIGNIFICANT GAINS FOR ANIMALS, AS WE SAW in chapter 4, worldwide meat consumption is at its highest level ever. To try to change this, some animal advocates, working with people seeking to protect the environment, are looking to alternative meat-like protein foods to compete with meat and save both animals and the planet. In 2021 Bloomberg, the financial data service, reported that sales of plant-based meat and milk products reached $29 billion in 2020 and are expected to grow to $162 billion by 2030. That's a rapid rate of growth, although even then, it would still be only 7.7 percent of the global protein market.[49] Such products already existed when I became a vegetarian more than fifty years ago, but at that time they were available only in the niche market catering to vegetarians. In contrast, when both the Impossible Burger and Beyond Meat burgers hit the market in 2016, they immediately went on sale alongside meat in supermarkets and fast-food chains. The next step that plant-based meat alternatives need to take is to compete with meat on price.

In 1931, Winston Churchill wrote an essay imagining the world "Fifty Years Hence," in which he looked forward to the day when "we shall escape the absurdity of growing a whole chicken in order to eat the breast or wing, by growing these parts separately under a suitable medium."[50] If Churchill had titled his essay "A Century Hence," his prophecy would have had a good chance of coming true, because progress is being made with cultivating animal cells in bioreactors to produce meat without breeding, raising, and killing animals. In 2021, Singapore licensed the sale of cellular chicken, but the product was more like a chicken nugget than chicken breast or wing. By 2022, nearly $2 billion had been invested in companies based in the United States, Europe, China, Japan, Israel, and South Africa, all seeking to produce cultivated meat or seafood, and several products are expected to reach the market in the next few years.[51] If they can compete with meat from animals for taste, texture, and price, they could be the key to spreading ethical eating worldwide and, eventually, ending factory farming. As Churchill said, growing whole animals in order to eat particular parts of the animal is an absurdity—and he didn't even consider what it is doing to our climate.

WHAT'S NEXT?

I AM OFTEN ASKED IF, WHEN I FIRST WROTE *Animal Liberation*, I expected it to have the success that it has had. The truth is that I didn't know what to expect. On the one hand, the core argument I was putting forward seemed so irrefutable, so undeniably right, that I thought everyone who read it would surely be convinced by it and would tell their friends to read it, and therefore everyone would stop eating meat and demand changes to our treatment of animals. On the other hand, in the 1970s, few people took issues concerning animals seriously. That speciesist attitude could have meant that the book would be ignored. If I succeeded in getting some attention, I was aware that the huge industries that exploit animals would fight against ideas that threatened their existence. Could rational and ethical arguments make headway against such powerful opposition? Alas, I thought, probably not.

What happened falls between these two opposing scenarios. Yes, there are more vegetarians and vegans than there were in 1975, and some of the reforms mentioned in this chapter have improved the lives of hundreds of millions of animals. On the other hand, there are now more animals suffering in laboratories and factory farms than ever before. We need much more radical changes than we have seen so far.

The animals themselves are incapable of demanding their own liberation, of protesting against their condition with rallies, votes, civil disobedience, or boycotts, or even of thanking those who advocate on their behalf. We humans have the power to continue to oppress other species forever, or until we make ourselves extinct. Will our tyranny continue, proving that morality counts for nothing when it clashes with self-interest, as many cynics have always said? Or will we rise to the challenge and prove our capacity for genuine altruism by ending our ruthless exploitation of the species over which we have power, not because we are forced to do so by rebels or terrorists, but because we recognize that our position is morally indefensible? I believe that this recognition will come, eventually, because over the past millennium we have made progress in expanding the sphere of those to whom we extend equal consideration. I do not know how long it will take for us to include nonhuman animals within this sphere, nor how many trillions of animals will continue to suffer until that happens. The way in which you and other readers respond to this book can shorten that time, and reduce that number.

ACKNOWLEDGMENTS

This book draws extensively from the 1975 edition of *Animal Liberation*, so I reiterate my thanks to those who opened my eyes to the indefensible treatment of animals that I supported for the first twenty-four years of my life: Richard and Mary Keshen, and Roslind and Stanley Godlovitch. In long conversations with these four—and particularly with Roslind, who had worked out her ethical position in considerable detail—I became convinced that by eating animals, I was participating in the systematic human oppression of other species. The central ideas of this book have their origin in those conversations that took place more than fifty years ago.

Reaching a theoretical conclusion is one thing; putting it into practice is another. Without the support and encouragement of Renata—my wife then and today—I might have continued to eat meat, though with a guilty conscience.

The idea of writing *Animal Liberation* arose from the enthusiastic response to my review of *Animals, Men and Morals* (edited by Stanley and Roslind Godlovitch and John Harris), which appeared in the *New York Review of Books* in April 1973. If Robert Silvers had not accepted my unsolicited discussion of an obscure book on an unfashionable topic, and then also edited and published the book that grew out of that review, it is unlikely that this book would exist. Others who made important contributions to the first edition are Richard Ryder, who coined the term *speciesism* and allowed me to draw on the materials he was assembling for his own book, *Victims of Science*, and Jim Mason, who helped to arrange my first visits to factory farms.

Animal Liberation was fully revised only once, for the second edition, published in 1990. For assistance with that update, I thank Lori Gruen, then a graduate student and now herself the author of several important books on animals and ethics.

For *Animal Liberation Now*, I was greatly aided by a grant from Open Philanthropy that made it possible for me to employ Sophie Kevany as a researcher. Sophie was assiduous in searching out information, much of which appears in chapters 2 and 3, although other parts of the book also benefited from her work. Sophie's research was in turn assisted by the many people and organizations listed below. I was also fortunate to have David Rosenbloom volunteer to do research on the environmental impact of meat production, on which I draw in chapter 4.

Yip Fai Tse provided comments on a draft of the entire text, as well as information on the treatment of animals in China. He encouraged me to expand the section on fish, and it is to Fai that I owe my knowledge of loaches and their treatment. I learned from Catia Faria, Oscar Horta, and, again, Yip Fai Tse, about the possibilities of reducing the suffering of wild animals, and from Paula Casal and Macarena Montes Franceschini about court cases in Latin America to free captive apes. Others who have influenced my thinking on an issue relevant to this book include Becca Franks, Guo Peng, Stevan Harnad, Jennifer Jacquet, Eva Kittay, Dale Jamieson, Arthur Shemitz, and Adam Lerner. Thanks to Kathy Robbins, my wonderful agent, as well as David Halpern and Janet Oshiro at The Robbins Office, for their advice at all stages of this project. Finally, I thank Sarah Haugen, my editor at HarperCollins, especially for her wise guidance on what to include and what to omit. Readers should be grateful that she reined me in when I found it difficult to resist providing even more details of the suffering inflicted on animals than the book now contains.

With special thanks to:

The Accountability Board: Josh Balk
American Anti-Vivisection Society: Jill Howard Church and Sue Leary
Animal Aid: Jessamy Korotoga
Animal Clock: Rick Gottlieb
Animal Ethics: Cyndi Rook
Animal Outlook: Piper Hoffman and Cheryl Leahy
Animal Welfare Institute: Marjorie Fishman, Dena Jones, and Gwendolen Reyes-Illg
Aquatic Life Institute: Christine Xu
Center for Alternatives to Animal Testing, Johns Hopkins Bloomberg School of Public Health: Kathrin Herrmann and Marty Stephens
Chennai Chapter of People For Animals: Shiranee Pereira
Conservative Animal Welfare Foundation: Lorraine Platt
Compassion in World Farming: Phil Brooke and Peter Stevenson
Counting Animals: Harish Sethu
Cruelty Free International: Katy Taylor
Fishcount.org: Alison Mood
Fish Welfare Initiative: Marco Cerquiera
Global Seafood Alliance: Steven Hedlund
The Humane League: Vicky Bond, Matthew Chalmers, and Mia Fernyhough

The Humane Society International: Wendy Higgins, Peter J. Li, Lindsay Marshall, Marcia Triunfol, Laura Viviani
The Humane Society of the United States: Bernard Unti, Josh Balk
Institute of Laboratory Animal Science: Kong Qi
Intergroup on the Welfare and Conservation of Animals: Andreas Erler, Luísa Ferreira Bastos
International Primate Protection League: Shirley McGreal
Laboratory Animal Science Welfare and Ethics Committee, China: Sun Deming
L214: Sébastien Arsac and Brigitte Gothière
Mercy for Animals: Leah Garcés
OECD-FAO Agricultural Outlook: Hubertus Gay
Our Honor: Crystal Heath
PETA: Alka Chandna, Frances Cheng, Kathy Guillermo, Magnolia Martinez, Emily Trunnell, Jennifer White, and the lab animals team
Procter & Gamble: Harald Schlatter
ProVeg: Charlotte Baker and Lara Pappers
Research Square: Michele Avissar-Whiting
Royal Society for the Prevention of Cruelty to Animals (UK): Penny Hawkins, Paul Littlefair, and Katherine Roe
U.S. Department of Agriculture: the communications and statistics teams
Vier-Pfoten (Four Paws), Austria: Teresa Pegger
Voices for Animals, Russia: Dinara Ageeva and Dmitry Tsvetkov
Vshine Animal Protection Association, China: Hong Mei Yu
WellBeing International: Andrew Rowan
White Coat Waste: Justin Goodman
World Animal Protection: George White

Thanks also to the following individuals: Jeremy Beckham, formerly of PETA; Susan Guynn, *Frederick News-Post*; Katherine Mills, McGill University; Sarah Pickering, Harvard University Animal Law and Policy Clinic; James Reynolds, Western University of Health Sciences; Alessandra Roncarati, University of Camerino; Sharon (who prefers to be thanked by that name only) and Michael Shuler, Cornell University.

To anyone I have inadvertently omitted, I offer my sincere apologies, and hope that the knowledge that you have made a difference is sufficient reward.

RECIPES

"This book begins with philosophy and ends with cookery." So began a review of the first edition of *Animal Liberation*. Why cookery? Because when we told our friends that we no longer ate meat, the first question they asked was: "What do you eat, then?" To answer that question, I added, at the end of the book, some vegetarian recipes. After dropping them from subsequent editions because I thought they were no longer necessary, I've been persuaded to again begin with philosophy and end with cookery. This time, though, I'm limiting myself to a few favorites—the recipes that I've most enjoyed cooking, eating, and sharing with family and friends.

My parents came to Australia in 1938, abandoning their beloved Vienna after the Nazi takeover of Austria. They brought with them the recipes they loved—almost entirely meat-based, except for the desserts and cakes. One of my childhood favorites, though, was pea soup with sippets. If you don't know what sippets are, you're in good company: Word's spellchecker is underlining it in red as I write. Read on and you will learn.

Austrian Pea Soup with Sippets

Serves 4

1 cup split peas
1 large onion, roughly chopped
2 to 3 cloves of garlic, finely minced
Olive oil
2 or 3 bay leaves
1 large carrot, roughly chopped
3 slices bread

Start with a cup of split peas, or if you are feeding a multitude, double the quantities. Soaking them overnight will reduce cooking time, but isn't really necessary. Then peel and chop one large onion, and (optionally) 2 or 3 cloves of garlic. Heat some oil in a large saucepan and add the garlic, which I put through a garlic crusher, but you can also chop finely. Let them brown, stirring occasionally. Then add the peas, drained if you soaked them, and stir them around for a minute or two with the onions.

Next add about 2 liters (4 pints) of water, with a teaspoon of salt and 2 to 3 bay leaves. Bring to a boil, then simmer. When the peas are starting to get soft, but not quite there, add the carrot. Continue to cook until the peas are soft, remove the bay leaves, and put the soup through a blender, or use a hand blender in the saucepan. The soup doesn't have to be really smooth; I leave a little texture in it.

Add salt and pepper to taste, and keep warm while you prepare the sippets, which are, to be honest, just the British version of croutons, but when freshly made are much more delicious than those boring dry things that come out of a packet. Take two or three slices of bread. I prefer a firm brown bread, and it can be stale, but any bread will do. Chop the bread into half-inch cubes. Heat some oil in a frying pan and fry the bread cubes until they are browned and slightly crisp. Serve immediately, putting them in a bowl so that they stay crisp and your diners can add them, a few at a time, to their soup as they eat it.

Austrian Lentil Soup

If you substitute brown or green lentils for the peas, you can use more or less the same recipe for Austrian Lentil Soup. I add two chopped potatoes when the carrot goes in. Again, I don't blend it all, so that some of the lentils still have their shape, and when the soup is ready, I stir in a teaspoon of lemon juice. Sippets are optional; they seem to go better with the pea soup, but are still good here.

Goulash

Serves 4

1 large onion, roughly chopped
2 garlic cloves, minced
3 to 4 small vegan sausages
3 to 4 large potatoes (russets work well)
3 bay leaves
Salt
1 to 2 teaspoons paprika
1 tablespoons flour (optional)

Goulash is Hungarian, not Austrian, but my parents were born before the First World War, when Austria was Austria-Hungary, so goulash was regularly on the table. Although goulash is often with meat, there are other kinds too. Every autumn my parents would go foraging for mushrooms, coming back with buckets full of saffron milk caps, a firm orange mushroom found under pines. It makes an excellent goulash, but unless you know how to distinguish edible mushrooms from deadly ones, please buy your mushrooms, or make this potato and sausage goulash, using, of course, some of the many excellent vegan sausages now widely available. A spicy sausage goes well in this, if that is to your taste. The key ingredient, though, is paprika. Standardly, this dish uses "sweet" paprika, which isn't really sweet, it just isn't hot. That's fine, but if your paprika has been sitting on your spice shelf for years, it will have lost its flavor. Get rid of it. Smoked paprika is worth trying, it gives a distinctive taste, and if you like your food spicy—or perhaps you didn't find any spicy vegan sausage—you can use hot paprika.

Start by frying a large chopped onion and some garlic in oil. Cut three or four small sausages into slices no more than half an inch thick, throw them in with the onion and garlic and let everything brown. Meanwhile peel three or four large potatoes, or an equivalent number of smaller ones, chop them into bite-size pieces, and throw them in too. Then add enough water to cover everything, together with 3 bay leaves, some salt, and one heaping teaspoon of paprika (unless you are using hot paprika, in which case start with less and then add more to your taste). Bring to a boil, then turn the heat to low and simmer for about 30 minutes, then adjust the seasoning. It may need a second teaspoon of paprika. The goulash is ready to serve when the potatoes are well-cooked and much of the water has been absorbed, or thickened into a sauce. If that hasn't happened, you can sprinkle a teaspoon of flour over the goulash and stir it in over low heat until it does.

Variations: You can add vegetables, such as red pepper, tomatoes, or zucchini. They will need less cooking time than the potatoes, so add them when the potatoes are starting to soften. You can also add herbs such as rosemary or thyme.

Renata, my wife, came to Australia as a young child from Poland. Her family's cooking was as meat-heavy as mine, but she has continued to make borscht, a soup that can vary from a clear liquid that has been strained to be free of the beetroot that gives it its color, to a much more hearty production that includes potatoes, carrots, cabbage, and, of course, beetroot. It's the latter that she makes.

Renata's Borscht

Serves 4

2 to 3 cloves garlic, chopped or crushed
1 large onion, roughly chopped
1 pound beetroots, scrubbed and cubed into 1-inch cubes
1 pound potatoes, peeled
Salt
Pepper
1 teaspoon sweet paprika
2 to 3 carrots, chopped (around 1 cup)
1 cup celery, chopped
1 cup cabbage, chopped
1 15-ounce can diced tomatoes
1 tablespoon tomato paste
Lemon juice to taste

Fry 2 or 3 cloves of crushed or chopped garlic and a large chopped onion in a little oil. Then add a pound of scrubbed beetroots chopped into 1-inch cubes. Add plenty of water—it has to cover the beetroots as well as the potatoes and vegetables still to come—salt, black pepper, a teaspoon of sweet paprika, and a bay leaf, and bring to a boil. Allow to simmer for 10 minutes, then add potatoes and carrots and simmer for another 5 minutes, or until the beetroot and potatoes are starting to soften just a little, but still need some more cooking. Add a cup of chopped celery, a cup of sliced cabbage, a can of chopped tomatoes, and a tablespoon of tomato paste, and continue to cook for another 5 minutes, or until the potatoes and beetroot are ready to eat. Adjust the seasoning and add a teaspoon of lemon juice, enough to give it a sweet-sour taste. Served with some good rye bread, this is a meal in itself. It's also one of those soups that gets even better on the second day, so you may want to increase the quantities.

When Renata and I stopped eating meat, we ate more Italian food, because there are many forms of pasta and pizza without meat that we already knew and liked. I won't give recipes here, because they are well-known. We also started experimenting with cuisines in which meat is less central than it is in Europe. India is home to more vegetarians than any other country in the world, and the first dish we made that we had never cooked before was dal, the lentil curry eaten daily by hundreds of millions of Indians. It's remained one of our favorite dishes. It starts the same way as most dishes I cook, and again, you can double the quantities to serve many, or just to freeze and have again on another occasion.

Dal

Serves 4

1 to 2 cloves garlic, minced or crushed
1 large onion

1 cup red lentils
Curry powder
Salt
Bay leaf
Cinnamon stick
1 14-ounce can chopped tomatoes
1 6-ounce can coconut milk
Lemon juice
Chopped greens (kale or spinach, optional)
Rice
Accompaniments (optional but highly desirable, available from Indian groceries or your supermarket's Asian section):

Mango chutney
Lime pickle
Poppadums

Chop 2 or 3 cloves of garlic or put them through a garlic crusher, and fry them in a little oil in a large saucepan. Add a large chopped onion and fry that. Then add a cup of small red lentils and a spoon or two of curry powder according to taste, with a pinch of salt. Stir that around in the pan for 2 or 3 minutes before adding 3 cups of water, a bay leaf, and a cinnamon stick. Bring to a boil, then turn very low and allow to simmer for about 20 minutes, stirring occasionally. Add a can of chopped tomatoes, and simmer a further 10 minutes. It should now be thick, and the lentils soft, but it should still be liquid enough to pour—just. Add some coconut milk and a little lemon juice. For variation, you can add some chopped spinach or kale a minute or two before you add the coconut milk. Serve over rice, with some lime pickle and mango chutney as accompaniments.

Indian restaurants will often serve dal with freshly chopped cucumbers in yoghurt, sometimes with ground coriander sprinkled over the top. Non-dairy yoghurts are now available in many supermarkets. Dal goes well with poppadums, which you can buy at supermarkets or at stores that sell Indian groceries. Put them, one at a time, briefly in very hot oil, deep enough to cover them. To test the temperature of the oil, break off a small piece of one poppadum and drop it in. If the oil is hot enough, it will instantly puff up. You can also do them under the grill, until they puff up—but take care, they burn easily.

Although there are fewer vegetarians in China than in India, there are many Chinese dishes that are vegan, or easily made vegan. A stir-fry has become one of my staples. For this, I use a wok, because you are less likely to spill the ingredients over the side when you toss them around, but any big frying pan will do. Here is the basic idea.

Tofu and Vegetable Stir-fry with Noodles or Rice

Serves 4

1 to 2 cloves garlic, minced or crushed
A similar quantity of ginger, finely chopped
4 scallions
14 ounces firm tofu
1 teaspoon soy sauce
1 tablespoon rice vinegar and/or Chinese cooking wine
Sesame oil, a few drops
1 cup each of any two or three of the following vegetables, chopped: broccoli, broccolini, cauliflower, bok choy, kale, cabbage, sweet red pepper, carrots, or bean sprouts
8 ounces Asian noodles or 1 cup rice

Prepare all the ingredients first, because once the frying is underway, it all goes very fast. Finely chop the garlic and ginger. Slice the scallions small, keeping the white parts separate from the green. Chop the vegetables more coarsely, and chop the tofu into 1-inch cubes.

If you are using rice, cook it separately, because you will serve it separately from the stir-fry. If your preference is noodles, prepare them ahead, as you are going to stir them into the wok at the end. Some Asian noodles need to be boiled for a few minutes and then drained, whereas others just need to have boiling water poured over them, left to stand for a few minutes, and then drained.

Frying the tofu is optional; it tastes better, but I don't like to consume too much oil, so sometimes I do it, and sometimes not. To fry it, put several tablespoons of oil in the wok, and when it is hot, add the tofu, stirring it so that it becomes golden brown and slightly crisp. Remove it from the oil and set aside. Now pour out most of the oil, leaving just a little in the bottom, and fry the garlic, ginger and the white parts of the scallions. Add a pinch of salt and half a teaspoon of sugar. When these ingredients are browned and smelling good, add the chopped vegetables. Put in first whatever will take longer to cook, for example broccoli or carrots. If you have not fried the tofu, add it now. Stir-fry until these veggies are cooked, then add the quicker-cooking vegetables, like leafy greens or bean sprouts, and if you have fried the tofu, add that too. If you have cooked noodles, they go in now, together with the green parts of the scallions. Stir through, making sure that the ingredients you cooked earlier and set aside are warm, sprinkle with a few drops of sesame oil and serve. If you are using rice instead of noodles, its best to put the rice on the plates or in the bowls and then add the stir-fry on top.

If you prefer plant-based meat-like products to tofu, you can use them too. There are endless variations.

Sichuan version: When frying the garlic and ginger, add some crushed chili, or use a teaspoon of Chinese chili oil—the dark red stuff available in Asian groceries that comes in a jar and needs to be stirred up. If you want to get really authentic, and your Asian grocery has it, try a tablespoon of Sichuan broad bean paste with chili oil, known in Chinese as doubanjiang; or for a different lip-numbing kind of heat, use some crushed Sichuan pepper. If some of those who will be eating your stir-fry like it spicy and others don't, omit the spices when cooking, and instead put chili oil on the table so that those who want it can stir it into their meal.

Sichuan Stir-fried Potatoes with Vinegar

I ate this dish at a Sichuan restaurant in New York, and it was so good—and such an unusual way of cooking potatoes, at least in Western countries—that I had to find a recipe and make it myself. Below is the recipe I like best, reprinted with the kind permission of Candy Hom, a Chinese American chef based in Atlanta, and the owner of Soupbelly LLC. It is available at https://soupbelly.com/2009/12/16/sichuan-stir-fried-potatoes-with-vinegar. If you are serving it as a main course for four people, I recommend doubling the quantities.

For this dish, the potatoes are cut to the size of matchsticks, soaked in salted water then stir-fried with Sichuan peppercorns, dried red chilies, sugar and vinegar, giving it a sweet, salty, and spicy flavor. The process of soaking the potatoes in salty water prevents them from turning brown, and the salt removes the excess starch. The texture should be slightly crunchy like a very al dente pasta.

Serves 4

3 medium size potatoes, cut into matchsticks
2 teaspoons cooking oil
3 large dried red chilies, chopped into pieces, or more small ones, according to how spicy you want it
1 teaspoon Sichuan peppercorns (crushed)
1 teaspoon sugar
3 teaspoons black vinegar (use Chinkiang vinegar if possible)
Salt to taste

Peel the potatoes and cut into matchsticks. Place matchsticks in a large bowl of cold, salted water (I put around 1 teaspoon of salt into the water) and soak (at least 10 minutes). Drain well.

In a wok on medium high heat, add oil, chilies, and Sichuan peppercorns. Heat until peppercorns are sizzling; add potatoes. Stir fry for around 4 minutes, add sugar and vinegar. Stir fry until al dente, add salt to taste. Serve hot.

A Chinese Cold Dish: Tofu and Kelp

I've already mentioned my mushroom foraging. When I'm at the beach, I look out for kelp, which is often there in huge quantities. It can be very thick and tough, but I take the thinner strands. Using kitchen scissors, I slice the kelp into strips, roughly an inch long and half an inch wide, and boil it for about 5 to 10 minutes, depending on how thick and tough it is. Drain it, and use it in the recipe below, or put it into a jar, cover with water and plenty of salt. It will then keep in the fridge for months, and you can use it in soups, or for this summer dish. If you lack foraging opportunities, you can buy dried kelp, and after soaking in water, use it like fresh kelp. This recipe would be good with other kinds of seaweed too.

Serves 4

1 block firm tofu
1/2 cup kelp, cut into strips
Ginger, chopped
2 to 3 radishes, sliced
1 to 2 scallion tops, sliced
1/2 teaspoon sugar
2 teaspoons soy sauce
2 teaspoons rice vinegar or Chinese vinegar
1 teaspoon sesame oil

Cut the tofu into half-inch cubes and place in a bowl. Add the kelp, some finely chopped fresh ginger, a couple of radishes, and the scallion tops, both sliced small, the sugar, soy sauce, vinegar, and sesame oil. If you want it spicy, add chili or chili oil to taste. Let it sit for 10 minutes so that the tofu marinates in the seasoning. (You may need to stir it a couple of times to make sure that all the tofu gets into the marinade.) It's now ready to eat.

Variations:

- Add finely chopped garlic (if raw garlic is too much for you, fry it lightly first, together with the bottoms of the scallions, sliced)
- Use finely chopped chives instead of scallions
- Add a tablespoon of chopped cilantro
- Add a handful of bean sprouts
- Lightly brown a handful of sesame seeds in a frying pan—you don't need oil—and throw them on top before serving

NOTES

CHAPTER 1: ALL ANIMALS ARE EQUAL . . .

1. John Stuart Mill refers to "Bentham's dictum" and describes it as "the first principle of the utilitarian scheme" and a principle of "perfect impartiality" in his *Utilitarianism,* ed. Katarzyna de Lazari-Radek and Peter Singer (New York: Norton Library, 2021, first published 1861) p. 81. Mill is paraphrasing a remark by Bentham in his *Rationale of Judicial Evidence, specially applied to English practice*, ed. J. S. Mill, 5 vols. (London, 1827), iv, 475 (book VIII, chapter XXIX); reprinted in *The Works of Jeremy Bentham*, ed. J. Bowring, Edinburgh, 1838–43, vii. 334. For the passage from Sidgwick, see *The Methods of Ethics*, 7th ed., (1907; reprint, London: Macmillan, 1963), p. 382. See also R. M. Hare, *Moral Thinking* (Oxford University Press, 1982), and for a brief account of the essential agreement on this issue between his and other positions, see R. M. Hare, "Rules of War and Moral Reasoning" in *Philosophy and Public Affairs* 1(2) (1972) 166–181. Rawls describes the "veil of ignorance" in *A Theory of Justice*, (Boston: Harvard University Press/Belknap Press, Cambridge, 1972).
2. I owe the term "speciesism" to Richard Ryder. It has become accepted in general use since the first edition of this book and now appears in The Oxford English Dictionary (OED). The word first appeared in the OED in 1986 and is listed, most recently, in the online OED, last modified in March 2018 and accessed August 13, 2022.
3. Jeremy Bentham, *Introduction to the Principles of Morals and Legislation* (1780), chapter 17.
4. Lucius Caviola, et al., "Humans first: Why people value animals less than humans," *Cognition* 225: 105139 (2022).
5. Stanley Benn, "Egalitarianism and Equal Consideration of Interests," in J. R. Pennock and J. W. Chapman, eds., *Nomos IX: Equality*, (New York, Atherton Press, 1967), p. 62ff; John Finnis, "The fragile case for euthanasia: a reply to John Harris," in J. Keown, ed., *Euthanasia examined: Ethical, clinical*

and legal perspectives (Cambridge University Press, 1995), pp. 46–55; Shelly Kagan, *How to Count Animals, more or less* (Oxford University Press, 2019).

6. See pp. 271ff.
7. Cambridge Declaration on Consciousness, https://fcmconference.org/img/CambridgeDeclarationOnConsciousness.pdf.
8. T. C., Danbury, et al., "Self-selection of the analgesic drug carprofen by lame broiler chickens," *Veterinary Record*, 146 (11): 307–11 (2000).
9. See Bernard Rollin, *The Unheeded Cry: Animal Consciousness, Animal Pain, and Science* (Oxford University Press, 1989).
10. Jane Goodall, *In the Shadow of Man*, (Boston: Houghton Mifflin, 1971), p. 225. Michael Peters makes a similar point in "Nature and Culture," in Stanley and Roslind Godlovitch and John Harris, eds., *Animals, Men and Morals* (New York: Taplinger, 1972). For examples of some of the inconsistencies in denials that creatures without language can feel pain, see Bernard Rollin, *The Unheeded Cry: Animal Consciousness, Animal Pain, and Science* (Oxford University Press, 1989).
11. Andrew Crump, et al., "Sentience in decapod crustaceans: A general framework and review of the evidence," *Animal Sentience* 32(1) (2022), www.wellbeingintlstudiesrepository.org/animsent/vol7/iss32/1/; see also Lynne Sneddon, "Pain in Aquatic Animals," *Journal of Experimental Biology* 218(7): 967–76 (2105).
12. See Helen Lambert, et al., "Given the Cold Shoulder: A Review of the Scientific Literature for Evidence of Reptile Sentience," *Animals (Basel)* 9: 10:821 (October 2019); and Helen Lambert, et al., "Frog in the well: A review of the scientific literature for evidence of amphibian sentience," *Applied Animal Behaviour Science* 247, article 105559 (2022).
13. Victoria Braithwaite, *Do Fish Feel Pain?* (Oxford University Press, 2010). The experiments I have summarized are described in chapters 3 and 4, and the quotation is on p. 113. The original source on hunting cooperation in fish is R. Bshary, et al., "Interspecific Communicative and Coordinated Hunting between Groupers and Giant Moray Eels in the Red Sea," *PLoS Biology* 4 (12): 431 (December 2006); see also L. Sneddon, "Pain in aquatic animals," *Journal of Experimental Biology* 218 (7): 967–76; L. Sneddon and C. Brown, "Mental Capacities of Fishes," in L. S. M. Johnson, A. Fenton, and A. Shriver, eds., *Neuroethics and Nonhuman Animals* (London: Springer Nature, 2020), 53–71. For a comparison of fish and primate cognitive abilities, see R. Bshary, et al., "Fish cognition: a primate's eye view," *Animal Cognition* 5 (1): 1–13 (2002).
14. Braithwaite, *Do Fish Feel Pain*, 48–49.
15. J. K. Finn, et al., "Defensive tool use in a coconut-carrying octopus," *Current Biology* 19 (23): R1069–70 (2009).
16. Peter Godfrey-Smith, *Other Minds: The Octopus, the Sea, and the Deep Origins of Consciousness* (New York: Farrar, Straus and Giroux, 2016).
17. Jonathan Birch, et al., *Review of the Evidence of Sentience in Cephalopod Molluscs and Decapod Crustaceans,* (London: LSE Consulting, November

2021), www.lse.ac.uk/business/consulting/reports/review-of-the-evidence-of-sentiences-in-cephalopod-molluscs-and-decapod-crustaceans; see also Andrew Crump, et al., "Sentience in decapod crustaceans: A general framework and review of the evidence," *Animal Sentience* 32 (1), www.wellbeingintlstudiesrepository.org/animsent/vol7/iss32/1 (2022). The UK Animal Welfare (Sentience) Act 2022 is available at www.legislation.gov.uk/ukpga/2022/22/enacted.

18. C. H. Eisemann, et al., "Do insects feel pain? A biological view," *Experientia* 40: 164–67 (February 1984).
19. Shelley Adamo, "Is it pain if it does not hurt? On the unlikelihood of insect pain," *Canadian Entomologist* 151 (6): 685–95 (2019).
20. Andrew Barron and Colin Klein, "What insects can tell us about the origins of consciousness," *Proceedings of the National Academy of Sciences* 113(18): 4900–908 (May 2016). For responses to this article, see Malte Schilling and Holk Cruse, "Avoid the hard problem: Employment of mental simulation for prediction is already a crucial step"; Shelley Adamo, "Consciousness explained or consciousness redefined?," and the reply by Colin Klein and Andrew Barron, "Crawling around the hard problem of consciousness," all in *Proceedings of the National Academy of Sciences* 113: E3811–15 (2016). See also Lars Chittka and Catherine Wilson, "Expanding Consciousness," *American Scientist* 107 (6): 364–69 (November–December 2019); and Matilda Gibbons, et al., "Descending control of nociception in insects?," *Proceedings of the Royal Society, B*, 289 (2022): 0599.
21. Kenny Torella, "Now is the best time to be alive (unless you're a farm animal)," Future Perfect, *Vox* September 12, 2022.
22. See Neil Gorsuch, *The Future of Assisted Suicide and Euthanasia* (New Jersey: Princeton University Press, 2006).
23. For details, see Gregory Pence, *Classic Cases in Bioethics*, 2nd ed. (New York: McGraw-Hill, 1995).
24. The philosopher Charles Camosy has argued, on several occasions when a guest speaker in my class on Practical Ethics at Princeton University, that the distinction between nonrational humans and nonhuman animals lies in the fact that humans have a rational nature, even if they are not actually rational. For discussion of a related view put forward by Shelly Kagan, see pp. 271ff.
25. See especially my *Rethinking Life and Death* (New York: St. Martin's Press, 1995) and *Practical Ethics*, 3rd ed. (Cambridge University Press, 2011).
26. See, for example, Dr. Irving Weissman, as quoted in Katherine Bishop, "From Shop to Lab to Farm, Animal Rights Battle is Felt," *New York Times*, January 14, 1989.

CHAPTER 2: TOOLS FOR RESEARCH . . .

1. Larry Carbone, "Estimating mouse and rat use in American laboratories by extrapolation from Animal Welfare Act-regulated species," *Scientific Reports*, 11–493 (2021) https://www.nature.com/articles/s41598–020–79961–0.

2. David Grimm, "How many mice and rats are used in US labs? Controversial study says more than 100 million," *Science*, January 12, 2021.
3. Zhiyan Consulting, "Heavyweight: Analysis of the licensed use and development prospects of China's laboratory animal industry in 2022," August 15, 2022, available at https://www.shangyexinzhi.com/article/5096882.html.
4. Katy Taylor and Laura Alvarez, "An Estimate of the Number of Animals Used for Scientific Purposes Worldwide in 2015," *Alternatives to Laboratory Animals* 47 196–213 (2019).
5. European Commission, *Summary Report on the statistics on the use of animals for scientific purposes in the Member States of the European Union and Norway in 2019*, Brussels, 15.7.2022, p. 3.
6. Mira van der Naald, et al., "Publication rate in preclinical research: a plea for preregistration," *BMJ Open Science* 4: e100051 (2020).
7. Shanghai Rankings, 2020 *Academic Ranking of World Universities*, www.shanghairanking.com/rankings/arwu/2020.
8. For the videos, see investigations.peta.org/nih-baby-monkey-experiments. For the statements by Goodall and Gluck, see investigations.peta.org/nih-baby-monkey-experiments/expert-statements. Barbara King expressed her view of the experiments in "Cruel Experiments on Infant Monkeys Still Happen All the Time—That Needs to Stop," *Scientific American* (June 1, 2015).
9. PETA, "NIH Ending Baby Monkey Experiments," www.peta.org/blog/nih-ends-baby-monkey-experiments/ (December 11, 2015).
10. Harry Harlow, et al., "Total Social Isolation In Monkeys," *Proceedings of the National Academy of Sciences* 54 (1): 90–97 (1965).
11. Harry Harlow and Stephen Suomi, "Induced Psychopathology in Monkeys," *Engineering and Science* 33 (6): 8–14 (1970).
12. John Bowlby, "Maternal Care and Mental Health," originally published in the *Bulletin of the World Health Organization* 3(3): 355–534 (March 1951).
13. Harlow and Soumi, "Induced psychopathology in monkeys."
14. Stephen Suomi and Harry Harlow, "Depressive behavior in young monkeys subjected to vertical chamber confinement," *Journal of Comparative and Physiological Psychology* 80 (1): 11–18 (1972).
15. Stephen Suomi and Harry Harlow, "Apparatus conceptualization for psychopathological research in monkeys, Instrumentation & Techniques," *Behavior Research Methods & Instrumentation* 1: 247–250 (January 1969).
16. Harry Harlow, et al., "Induction of psychological death in rhesus monkeys," *Journal of Autism and Development Disorders* (formerly *Journal of Autism and Childhood Schizophrenia*) 3: 299–307 (October 1973).
17. Gene Sackett, et al., "Social isolation rearing effects in monkeys vary with genotype," *Developmental Psychology* 17 (3): 313–18 (1981).
18. Deborah Snyder, et al., "Peer separation in infant chimpanzees, a pilot study," *Primates* 25: 78–88 (1984).
19. From *I, Candidate for Governor* (1935), in Susan Ratcliffe, ed., *Oxford Es-*

sential Quotations, 4th ed., www.oxfordreference.com/view/10.1093/acref/9780191826719.001.0001/acref-9780191826719 (2016).

20. Based on a search of NIH RePorter conducted by Sophie Kevany, February 2023, using the following search terms: Fiscal Year: 2020; Agency/Institute/Center: Nat'l Inst of Mental Health (NIMH); Funding: Yes; Text Search: "animal model," limit to: Project Terms.
21. Richard Solomon, et al., "Traumatic avoidance learning: the outcomes of several extinction procedures with dogs," *Journal of Abnormal and Social Psychology* 48 (2): 291–302 (1953).
22. Martin Seligman, et al., "Alleviation of Learned Helplessness in the Dog," *Journal of Abnormal Psychology* 73 (3, pt. 1): 256–62 (1968).
23. Seligman, "Alleviation of Learned Helplessness in the Dog."
24. Steven Maier and Martin Seligman, "Learned Helplessness at Fifty: Insights from Neuroscience," *Psychological Review* 123 (4): 349–67 (July 2016).
25. Gary Brown, et al., "Effect of Escapable versus Inescapable Shock on Avoidance Behavior in the Goldfish (Carassius Auratus)," *Psychological Reports* 57 (3 suppl): 1027–30 (1985).
26. Steven Maier, "Learned helplessness and animal models of depression," *Progress in Neuro-Psychopharmacology and Biological Psychiatry* 8 (3): 435–46 (1984).
27. Hielke Van Dijken, et al., "Inescapable footshocks induce progressive and long-lasting behavioural changes in male rats," *Physiology & Behavior* 51 (4): 787–94 (April 1992).
28. Bibiana Török, et al., "Modelling posttraumatic stress disorders in animals," *Progress in Neuro-Psychopharmacology and Biological Psychiatry* 90: 117–33 (March 2019).
29. Lei Zhang, et al., "Updates in PTSD Animal Models Characterization," in Firas H. Kobeissy, ed., *Psychiatric Disorders: Methods in Molecular Biology* (Totowa, NJ: Humana, 2012), 331–44.
30. Meghan Donovan, et al., "Anxiety-like behavior and neuropeptide receptor expression in male and female prairie voles: The effects of stress and social buffering," *Behavioral Brain Research* 342: 70–78 (April 2018). See also Claudia Lieberwirth and Zuoxin Wang, "The neurobiology of pair bond formation, bond disruption, and social buffering," *Current Opinion in Neurobiology* 40: 8–13 (October 2016); Adam Smith and Zoxin Wang, "Hypothalamic oxytocin mediates social buffering of the stress response," *Biological Psychiatry* 76 (4): 281–88 (August 2014).
31. Teng Teng, et al., "Chronic unpredictable mild stress produces depressive-like behavior, hypercortisolemia, and metabolic dysfunction in adolescent cynomolgus monkeys," *Translational Psychiatry* 11 (9) (January 2021).
32. Weixin Yan, et al., "fMRI analysis of MCP-1 induced prefrontal cortex neuronal dysfunction in depressive cynomolgus monkeys" (preprint, available from Research Square, www.researchsquare.com/article/rs-10408/v2, accessed 9 August 2021). The website states that "Research Square has withdrawn this preprint due to ethical concerns."

33. Yin Y-Y, et al., "The Faster-Onset Antidepressant Effects of Hypidone Hydrochloride (YL-0919) in Monkeys Subjected to Chronic Unpredictable Stress," *Frontiers in Pharmacology* 11 article 586879 (November 2020).
34. Letters from Emily Trunnell to Julio Licinio, editor in chief, *Translational Psychiatry*; to Michelle Avissar-Whiting, editor in chief, Research Square; and to Heike Wulff, field editor in chief, *Frontiers in Pharmacology*, all dated February 4, 2021.
35. *Summary Report on the statistics on the use of animals for scientific purposes in the Member States of the European Union and Norway in 2019*, p.31.
36. *Summary Report on the statistics on the use of animals for scientific purposes in the Member States of the European Union and Norway in 2019*, p.8. Note that in the EU, some statistics are reported in terms of "uses of animals" rather than numbers of animals used. Some animals may be used more than once, but generally speaking the figure is not very different from the number of animals used.
37. Home Office, United Kingdom, "*Annual Statistics of Scientific Procedures on Living Animals, Great Britain, 2019*," and associated tables, ordered by the House of Commons to be printed July 16, 2020.
38. *Summary Report on the statistics on the use of animals for scientific purposes in the Member States of the European Union and Norway in 2019*, p.6.
39. Jeffrey Aronson and Richard Green, "Me-too pharmaceutical products: History, definitions, examples, and relevance to drug shortages and essential medicines lists," *British Journal of Clinical Pharmacology* (86) 11: 2114–22 (November 2020).
40. U.S. Congress, Office of Technology Assessment, *Alternatives to Animal Use in Research, Testing and Education* (Washington, D.C.: U.S. Government Printing Office, 1986), 168.
41. Regina Arantes-Rodrigues, et al., "The effects of repeated oral gavage on the health of male CD-1 mice," *Lab Animal* 41 (5): 129–134 (May 2012).
42. The scale is cited in K. J. Olson et al., Toxicological Properties of Several Commercially Available Surfactants, *Journal of the Society of Cosmetic Chemists* 13 (9): 470, library.scconline.org/v013n09/35 (May 1962).
43. Cristine Russell and Penny Chorlton, "Eli Lilly Removes Arthritis Drug From Market," *Washington Post*, August 5, 1982; Morton Mintz, "Ga. Jury Awards $6m in Oraflex Deaths," *Washington Post* November 22, 1983.
44. Matthew Herper, "David Graham on the Vioxx Verdict," *Forbes* August 19, 2005.
45. John Gartner, "Vioxx Suit Faults Animal Tests," *Wired* July 22, 2005.
46. U.S. Department of Justice, "U.S. Pharmaceutical Company Merck Sharp and Dohme to pay nearly one billion dollars Over Promotion of Vioxx," www.justice.gov/opa/pr/us-pharmaceutical-company-merck-sharp-dohme-pay-nearly-one-billion-dollars-over-promotion (November 22, 2011).
47. G. E. Paget, ed., *Methods in Toxicology* (Oxford: Blackwell Scientific, 1970), 132.

48. G. F. Somers, *Quantitative Method in Human Pharmacology and Therapeutics* (Elmsford, NY: Pergamon Press, 1959), quoted by Richard Ryder, *Victims of Science* (Fontwell, Sussex: Centaur Press/State Mutual Book, 1983), 153.
49. Syndicated article appearing in *West County Times* (California) January 17, 1988.
50. Michael Bracken, "Why animal studies are often poor predictors of human reactions to exposure," *Journal of the Royal Society of Medicine* 102 (3): 120–22 (2009).
51. Michael Schuler, "Organ-, body- and disease-on-a-chip," *Lab on a Chip* 17: 2345–46 (2017).
52. *New York Times*, April 15, 1980.
53. For the quote from Roger Shelley and other details, see Peter Singer, *Ethics into Action: Learning from a Tube of Toothpaste* (Lanham, MD: Rowman and Littlefield, 2019).
54. "Avon Validates Draize Substitute," News Release, Avon Products, New York, April 5, 1989; "Avon Announces Permanent End to Animal Testing," News Release, Avon Products, New York, June 22, 1989.
55. "Industry Toxicologists Keen on Reducing Animal Use," *Science* 236 (4799): 252 (April 17, 1987).
56. Barnaby J. Feder, "Beyond White Rats and Rabbits," *New York Times*, February 28, 1988; see also Constance Holden, "Much Work But Slow Going on Alternatives to Draize Test," *Science* 242 (4876): 185–86 (October 14, 1985).
57. Coalition to Abolish the LD50, "Animal Rights Coalitions Coordinator's Report '83" (1984). Animal Rights Coalitions. 4. https://www.wellbeingintl studiesrepository.org/anircoa/4
58. Paul Cotton, "Animals and Science Benefit From 'Replace, Reduce, Refine' Effort," *Journal of the American Medical Association* 270 (24): 2905–907 (1993).
59. The figures on the number of animals required to meet OECD guidelines are from John Doe and Philip Botham, "Chemicals and Pesticides: A Long Way to Go," in Michael Balls, Robert Combes, and Andrew Worth, eds., *The History of Alternative Test Methods in Toxicology*, (London: Academic Press, 2019), 177–84.
60. U.S. Environmental Protection Agency, "Revised Final Health Effects Test Guidelines; Acute Toxicity Testing-Background and Acute Oral Toxicity; Notice of Availability," *Federal Register*, Notice 67, 77064–65 (December 16, 2002).
61. Nicholas St. Fleur, "N.I.H. to end backing for invasive research on chimps," *New York Times*, November 19, 2015.
62. David Grimm, "U.S. EPA to eliminate all mammal testing by 2035," *Science* September 10, 2019; Mihir Zaveri, Mariel Padilla, and Jaclyn Peiser, "EPA says it will drastically reduce animal testing," *New York Times*, September 10, 2019.

63. See Doe and Botham, "Chemicals and Pesticides: A Long Way to Go."
64. *Summary Report on the statistics on the use of animals for scientific purposes in the Member States of the European Union and Norway in 2019*, p.45.
65. See Table 7.4 of the data used in the UK Home Office, "Annual Statistics of Scientific Procedures on Living Animals, Great Britain: 2019," ordered by the House of Commons to be printed July 16, 2020, annual-statistics -scientific-procedures-living-animals-2019-tables.
66. Leandro Texeira and Richard Dubielzig, "Eye," in Wanda M. Haschek, et al., eds., *Haschek and Rousseaux's Handbook of Toxicologic Pathology* (Amsterdam: lsevier Science & Technology, 2013), 2128.
67. Francois Busquet, et al, "New European Union Statistics on Laboratory Animal Use: What Really Counts!" *Altex* 37(2): 167-186 (2020), at p.179; and for the 2019 figure, *Summary Report on the statistics on the use of animals for scientific purposes in the Member States of the European Union and Norway in 2019*, p.44.
68. Thomas Hartung, et al., "New European Union Statistics on Laboratory Animal Use–What Really Counts!" *Altex: Alternativen zu Tierexperimenten* 37 (2): 167–86 (March 2020).
69. OECD (2021), Test No. 405: Acute Eye Irritation/Corrosion, OECD Guidelines for the Testing of Chemicals, Section 4, OECD Publishing, Paris, doi.org/10.1787/9789264185333-en.
70. Magnus Gregersen, "Shock," *Annual Review of Physiology* 8: 335–54 (March 1946).
71. Kirtland Hobler and Rudolph Napodano, "Tolerance of swine to acute blood volume deficits," *Journal of Trauma: Injury, Infection, and Critical Care* 14 (8): 716–18 (August 1974).
72. A. Fülöp, et al., "Experimental Models of Hemorrhagic Shock: A Review," *European Surgical Research* 50 (2): 57–70 (June 2013).
73. L. F. McNicholas, et al., "Physical dependence on diazepam and lorazepam in the dog," *Journal of Pharmacology and Experimental Therapeutics* 226 (3): 783–89 (September 1983).
74. Ronald Siegel, "LSD-induced effects in elephants: Comparisons with musth behavior," *Bulletin of the Psychonomic Society* 22 (1): 53–56 (1984).
75. Michiko Okamoto, et al., "Withdrawal characteristics following chronic pentobarbital dosing in cat," *European Journal of Pharmacology* 40 (1): 107–19 (1976).
76. "TSU Shuts Down Cornell Cat Lab," *Newsweek* December 26, 1988, 50; "TSU Shuts Down Cat Lab at Cornell," *Animals' Agenda*, 22–23, newspaper .animalpeopleforum.org/wp-content/uploads/2018/08/AnimalsAgenda March1989.pdf (March 1989).
77. Kathryn Harper, et al., "Age-related differences in anxiety-like behavior and amygdalar CCL2 responsiveness to stress following alcohol withdrawal in male Wistar rats," *Psychopharmacology (Berl)* 234 (1): 79–88 (January 2017).
78. Matt Field and Inge Kersbergen, "Are animal models of addiction useful," *Addiction* 115 (1): 6–12 (July 2019).

79. Stanley Milgram, *Obedience to Authority* (New York Harper & Row, 1974); for a recent replication, see Dariusz Doliński, et al., "Would You Deliver an Electric Shock in 2015? Obedience in the Experimental Paradigm Developed by Stanley Milgram in the 50 Years Following the Original Studies," *Social Psychological and Personality Science* 8 (8): 927–33 (2017).
80. Blick, D. W., et al., "Primate equilibrium performance following soman exposure: Effects of repeated daily exposures to low soman doses," U.S. Air Force School of Aerospace Medicine Report No. USAFSAM-TR-87–19, Brooks AFB TX, October 1987.
81. U.S. Air Force, School of Aerospace Medicine, Report No USAFSAM-TR-87–19, 6.
82. Donald J. Barnes, "A Matter of Change," in Peter Singer, ed., *In Defense of Animals* (Oxford: Blackwell, 1985).
83. Barnes, "A Matter of Change," 160–66.
84. Steven Pinker, *The Better Angels of Our Nature* (New York: Viking, 2011), 455–56.
85. United Action for Animals, *The Death Sciences in Veterinary Research and Education* (New York, undated), iii.
86. Stan Wayman, "Concentration Camps for Dogs," *Life* February 3, 1966; Coles Phinizy, "The Lost Pets that Stray to the Labs," *Sports Illustrated* November 29, 1965. For more information, see Christine Stevens, "Laboratory Animal Welfare," in *Animals and Their Legal Rights* (Washington, D.C.: Animal Welfare Institute, 1990), 66–111.
87. *Journal of the American Veterinary Medical Association* 163 (9) (November 1, 1973).
88. Dominique Potvin, "Altruism in birds? Magpies have outwitted scientists by helping each other remove tracking devices," *Conversation* 21, theconversation.com/altruism-in-birds-magpies-have-outwitted-scientists-by-helping-each-other-remove-tracking-devices-175246 (February 2022); for the scientific publications reporting this, see Joel Crampton, et al., "Australian Magpies *Gymnorhina tibicen* cooperate to remove tracking devices," *Australian Field Ornithology* 39: 7–11 (2022).
89. Email from Dominique Potvin to the author, February 26, 2022.
90. Harry Harlow, editorial, "Fundamental Principles for Preparing Psychology Journal Articles," *Journal of Comparative and Physiological Psychology* 55 (6): 893–96 (1962).
91. Vanessa von Kortzfleisch, et al., "Improving reproducibility in animal research by splitting the study population into several 'mini-experiments,'" *Nature, Scientific Reports* 10, article 16579 (October 2020).
92. C. Glenn Begley and John P. A. Ioannides, "Reproducibility in Science: Improving the Standard for Basic and Preclinical Research," *Circulation Research* 116 (1): 116–26 (2015).
93. Leonard Freedman, et al., "The Economics of Reproducibility in Preclinical Research," *PLoS Biology* 13 (6): e1002165 (June 2015).

94. Marlene Cimons, et al., "Cancer Drugs Face Long Road from Mice to Men," *Los Angeles Times* May 6,1998. I owe this quotation and the one that follows from Elias Zerhouni to Jim Keen, "Wasted money in United States Biomedical and Agricultural Animal Research," in Kathrin Hermann and Kimberley Jayne, *Animal Experimentation: Working Towards a Paradigm Change* (Leiden, The Netherlands: Brill, 2019), 244–72.
95. Junhee Seok, et al., "Genomic responses in mouse models poorly mimic human inflammatory diseases," *Proceedings of the National Academy of Sciences* 110 (9): 3507–12 (2013).
96. Rich McManus, "Ex-Director Zerhouni Surveys Value of NIH Research," *NIH Record* LXV (13): 4 (June 21, 2013).
97. A report by the Medicines Discovery Catapult and the Bioindustry Association, *State of the Discovery Nation 2018,* md.catapult.org.uk/resources/report-state-of-the-discovery-nation-2018/.
98. *Birmingham News* (AL), February 12, 1988.
99. "The Price of Knowledge," broadcast in New York, December 12, 1974, WNET/13, transcript supplied courtesy of WNET/13.
100. Quoted in U.S. Congress, Office of Technology Assessment, *Alternatives to Animal Use in Research, Testing and Education*, 277.
101. Teng Teng, Carol Shively, et al., "Chronic unpredictable mild stress produces depressive-like behavior, hypercortisolemia, and metabolic dysfunction in adolescent cynomolgus monkeys," *Translational Psychiatry* 11, article 9 (2021).
102. Yin Y-Y, et al. "The Faster-Onset Antidepressant Effects of Hypidone Hydrochloride."
103. *National Health and Medical Research Council, Australian Code for the Care and Use of Animals for Scientific Purposes* (first published 1985, current edition 2013), available at www.nhmrc.gov.au/about-us/publications/australian-code-care-and-use-animals-scientific-purposes. For discussion, see Aaron Timoshenko, et al., "Australian Regulation of Animal Use in Science and Education: A Critical Appraisal," *ILAR Journal* 57(3): 324–32 (2016).
104. OTA, *Alternatives to Animal Use in Research, Testing, and Education*, 377.
105. OTA, *Alternatives to Animal Use in Research, Testing, and Education*, 286.
106. OTA, *Alternatives to Animal Use in Research, Testing, and Education*, 287.
107. Animal and Plant Health Inspection Service, U.S. Department of Agriculture, *Animal Welfare Act and Animal Welfare Regulations*, May 2019, p. 6. For an account of the events described, see American Anti-Vivisection Society, "Birds, Rats and Mice," aavs.org/our-work/campaigns/birds-mice-rats/, accessed 11 August 2022.
108. Pinker, *Better Angels of Our Nature*, 455–56.
109. Emily Trunnell, email to the author, June 9, 2021, and "Test Subjects," a film directed and produced by Alex Lockwood in 2020 and available at testsubjectsfilm.com.

110. For an account of the Willowbrook experiment, see en.wikipedia.org/wiki/Willowbrook_State_School#Hepatitis_studies. The "most unethical" comment is by Maurice Hilleman and is quoted in Paul Offit, *Vaccinated: One Man's Quest to Defeat the World's Deadliest Diseases* (Washington, D.C.: Smithsonian Books/Collins, 2007), 27.
111. See Richard Yetter Chappell and Peter Singer, "Pandemic ethics: the case for risky research," *Research Ethics* 16 (3–4): 1–8 (2020); Peter Singer and Isaac Martinez, "The Case for Human Covid-19 Challenge Trials," *Project Syndicate* (August 5, 2020).
112. R. J. Lifton, *The Nazi Doctors* (New York: Basic Books, 1986).
113. I. B. Singer, *Enemies: A Love Story* (New York: Farrar, Straus and Giroux, 1972), 257.
114. See James Jones, *Bad Blood: The Tuskegee Syphilis Experiment* (New York: Free Press, 1981).
115. Sandra Coney, *The Unfortunate Experiment* (Auckland, Penguin Books, 1988).
116. The documentary "Monkeys, Rats and Me: Animal Testing" was first screened by the BBC on November 27, 2006; the review article is by Laure Pycroft, John Stein, and Tipu Aziz, "Deep brain stimulation: an overview of history, methods, and future developments," *Brain and Neuroscience Advances* 2: 1–6 (2018).
117. Thomas McKeown, *The Role of Medicine: Dream, Mirage or Nemesis?* (Oxford: Blackwell, 1979).
118. David St. George, "Life Expectancy, Truth, and the ABPI," *Lancet* 328 (8502), August 9, 1986, 346.
119. J. B. McKinlay, S. M. McKinley, and R. Beaglehole, "Trends in Death and Disease and the Contribution of Medical Measures," in H. E. Freeman and S. Levine, eds., *Handbook of Medical Sociology* (Englewood Cliffs, NJ: Prentice Hall, 1988), 16.
120. See William Paton, *Man and Mouse* (Oxford University Press, 1984); Andrew Rowan, *Of Mice, Models and Men: A Critical Evaluation of Animal Research*, Albany: State University of New York Press, 1984, chapter 12; Michael DeBakey, "Medical Advances Resulting From Animal Research," in J. Archibald J. Ditchfield and H. Rowsell, eds., *The Contribution of Laboratory Animal Science to the Welfare of Man and Animals: Past, Present and Future* (New York: Gustav Fischer Verlag, 1985); OTA, *Alternatives to Animal Use in Research, Testing, and Education*, chapter 5; and National Research Council, *Use of Animals in Biomedical and Behavioral Research* (Washington, D.C.: National Academy Press, 1988), chapter 3.
121. See, for example, Ray Greek and Niall Shanks, *FAQs About Animal Research* (Lanham, MD: University Press of America, 2009); Robert Sharpe, *The Cruel Deception* (New York: HarperCollins, 1988).
122. UNICEF, "Under-five Mortality," data.unicef.org/topic/child-survival/under-five-mortality, accessed August 11, 2022. See also World Health Organization,

"Children: improving survival and wellbeing," www.who.int/news-room/fact-sheets/detail/children-reducing-mortality (September 8, 2020).

123. Eurogroup for Animals, "A win for animals! The European Parliament votes in favor of a comprehensive plan to phase-out experiments on animals," www.eurogroupforanimals.org/news/win-animals-european-parliament-votes-favour-comprehensive-plan-phase-out-experiments-animals (September 16, 2021).

CHAPTER 3: DOWN ON THE FACTORY FARM . . .

1. Angela Baysinger, et al., "A case study of ventilation shutdown with the addition of high temperature and humidity for depopulation of pigs," *Journal of the American Veterinary Medical Association* 259 (4): 415–24 (August 2021); see also Gwendolen Reyes-Illg, et al., "The rise of heatstroke as a method of depopulating pigs and poultry: Implications for the US veterinary profession," *Animals* 13: 140 (2023).
2. Figures for the number of land animals slaughtered in the United States are from the U.S. Department of Agriculture, figures from 2021. See USDA National Agriculture Statistics Service, *Poultry Slaughter*, January 24, 2022, p. 2; and USDA National Agriculture Statistics Service, *Livestock Slaughter*, January 20, 2022, p. 13.
3. Figures taken by selecting the relevant items from www.fao.org/faostat/en/#home (accessed January 9, 2023).
4. See "meat" in the *Online Etymological Dictionary*, www.etymonline.com/word/meat.
5. Fact Sheet: The Biden-Harris Plan for a Fairer, More Competitive and More Resilient Meat and Poultry Supply Chain, White House, January 3, 2022, www.whitehouse.gov/briefing-room/statements-releases/2022/01/03/fact-sheet-the-biden-harris-action-plan-for-a-fairer-more-competitive-and-more-resilient-meat-and-poultry-supply-chain/.
6. The calculation of percentages is based on figures from the U.S. Department of Agriculture, National Agricultural Statistics Service, "Chickens and Eggs: 2021 Summary," February 2022; and Terrence O'Keefe, "The largest US egg-producing companies of 2022," Wattagent January 19, 2022, https://www.wattagnet.com/articles/44099-the-largest-us-egg-producing-companies-of-2022.
7. Lucy King, Adam Westbrook, and Jonah Kessel, "See the True Cost of Your Cheap Chicken," *New York Times*, February 10, 2022, www.nytimes.com/2022/02/10/opinion/factory-farming-chicken.html.
8. Ruth Harrison, *Animal Machines* (London: Vincent Stuart, 1964), 3.
9. As this book goes to press, the North American Meat Institute's attempt to block California's ban on the sale of products from animals kept in conditions that do not comply with California's law is before the U.S. Supreme Court. For details see "California's Proposition 12, 2018," Wikipedia, https://en.wikipedia.org/wiki/2018_California_Proposition_12.

10. U.S. Department of Agriculture, *Broiler Market News Report*, January 2023.
11. Konrad Lorenz, *King Solomon's Ring* (London: Methuen and Company, 1964), 147.
12. See, for example, Brendan Graaf, "Lighting Considerations for Commercial Broiler" (Siloam Springs, AK: Cobb-Vantress, Inc, 201).
13. National Chicken Council, "U.S. Broiler Performance," www.nationalchickencouncil.org/about-the-industry/statistics/u-s-broiler-performance/, accessed February 2022; Marie-Laure Augère-Granier, *The EU Poultry Meat and Egg Sector*, European Parliamentary Research Service PE644–195, November 2019.
14. National Chicken Council, U.S. Broiler Performance, www.nationalchickencouncil.org/about-the-industry/statistics/u-s-broiler-performance/, accessed February 2022.
15. Stéphane Bergeron, Emmanuelle Pouliot, and Maurice Doyon, "Commercial Poultry Production Stocking Density Influence on Bird Health and Performance Indicators," *Animals* 10(8): 1253 (2020).
16. Eurogroup for Animals, *The Welfare of Broiler Chickens in the EU* (Brussels, 2020), 12.
17. M. O. North and Bell D. D., *Commercial Chicken Production Manual*, 4th ed. (New York: Van Nostrand Reinhold, 1990), 456.
18. King, Westbrook, and Kessel, "See the True Cost of Your Cheap Chicken."
19. Casey W. Ritz, Brian D. Fairchild, and Michael P. Lacy, "Litter Quality and Broiler Performance," Poultry Site, August 22, 2005, www.thepoultrysite.com/articles/litter-quality-and-broiler-performance.
20. U.S. Department of Agriculture Yearbook for 1970, xxxiii.
21. Katy Mumaw, "Contract Chickens, Get an Inside Look," *Farm and Dairy* April 26, 2018, www.farmanddairy.com/uncategorized/contract-chickens-get-an-inside-look/483606.html.
22. Chris Harris, "Broiler Production and Management," Poultry Site, April 24, 2004, www.thepoultrysite.com/articles/broiler-production-and-management.
23. National Chicken Council, U.S. Broiler Performance, www.nationalchickencouncil.org/about-the-industry/statistics/u-s-broiler-performance/, accessed February 2022.
24. U.S. Department of Agriculture, National Agricultural Statistics Service, *Poultry Slaughter*, 2021 Summary, February 2022.
25. Jean Sander, "Sudden Death Syndrome in Broiler Chickens," *MSD Veterinary Manual*, November 2019, www.msdvetmanual.com/poultry/sudden-death-syndrome-of-broiler-chickens/sudden-death-syndrome-of-broiler-chickens.
26. Fabian Brockotter, "Controlling Sudden Death Syndrome via Feed Strategies," *Poultry World*, May 1, 2020.
27. R. F. Wideman, D. Rhoads, G. Erf, and N. Anthony, "Pulmonary arterial hypertension (ascites syndrome) in broilers: A review," *Poultry Science* 92 (1): 64–83 (2013).
28. S. C. Kestin, T. G. Knowles, A. E. Tinch, and N. G. Gregory, "Prevalence

of Leg Weakness in Broiler Chickens and its Relationship with Genotype," *Veterinary Record* 131 (9): 190–194 (August 1992).

29. Professor Webster's comment is from *The Guardian*, October 14, 1991.
30. John Webster, *Animal Welfare: A Cool Eye Towards Eden* (Oxford: Blackwell Science, 1995), p. 156.
31. "Pilgrim's Shame: Chickens Buried Alive," *Animal Outlook,* animaloutlook.org/investigations/pilgrims/. The total revenue of the U.S. poultry industry in 2021 was $46.1 billion, of which 68 percent was from broilers. See U.S. Department of Agriculture, National Agricultural Statistics Service, "Poultry—Production and Value, 2021, Summary," April 2022.
32. Media release from Animal Law Italia and Equalia, February 16, 2022.
33. Arnaud van Wettere, "Noninfectious skeletal disorders in poultry broilers," *Merck Veterinary Manual* (last modified 2022), www.merckvetmanual.com/poultry/disorders-of-the-skeletal-system-in-poultry/noninfectious-skeletal-disorders-in-poultry-broilers.
34. G. T. Tabler and A. M. Mendenhall, "Broiler Nutrition, Feed Intake and Grower Economics," *Avian Advice* 5 (4): 9 (Winter 2003).
35. J. Mench, "Broiler breeders: feed restriction and welfare, *World's Poultry Science Journal* 58: 23–29 (2002).
36. K. M. Wilson, et al., "Impact of Skip-a-Day and Every-Day Feeding Programs for Broiler Breeder Pullets on the Recovery of Salmonella and Campylobacter following challenge," *Poultry Science* 97 (8): 2775–84 (August 1, 2018).
37. A. Arrazola, et al., "The effect of alternative feeding strategies on the feeding motivation of broiler breeder pullets," *Animal* 14 (10): 2150–58 (October 2020).
38. National Chicken Council, "National Chicken Council Animal Welfare Guidelines and Audit Checklist," June 2017, www.nationalchickencouncil.org/wp-content/uploads/2017/07/NCC-Welfare-Guidelines_Broiler Breeders.pdf.
39. John Vidal, *McLibel: Burger Culture on Trial* (London: Pan Books, 1997), p. 311.
40. See "Welcome to Peer System," www.peersystem.nl/en/. For the video, see www.youtube.com/watch?v=8mFP4isNuMY (accessed 26 April 2022).
41. For information on the welfare aspects of chicken slaughter and a comparison of different methods of stunning, see Charlotte Berg and Mohan Raj, "A Review of Different Stunning Methods for Poultry–Animal Welfare Aspects," *Animals* 5 (4): 1207–19 (2015).
42. National Chicken Council, "National Chicken Council Animal Care Guidelines Certified by Independent Audit Certification Organization," July 10, 2018, www.nationalchickencouncil.org/national-chicken-council-animal-care-guidelines-certified-by-independent-audit-certification-organization/.
43. The video taken at the uninvited inspection is available at www.youtube.com/watch?v=b6A1kWnEfqk. The quotations from spokespeople for Moy

Park and Niko Omilana are from Andrew Wasley, "KFC faces backlash over 'misleading' portrayal of UK chicken farming," *Guardian*, April 12, 2022.

44. Edgar Oviedo-Rondón, "Predisposing Factors that Affect Walking Ability in Turkeys and Broilers," February 1, 2009, www.thepoultrysite.com/articles/predisposing-factors-that-affect-walking-ability-in-turkeys-and-broilers# (accessed 26 April 2022); V. Allain, et al., "Prevalence of skin lesions in turkeys at slaughter," *British Poultry Science* 54 (1): 33–41 (2013); Hybrid, "What are breast blisters and buttons," www.hybridturkeys.com/en/news/preventing-breast-blisters-and-buttons/ (accessed 26 April 2022).
45. This account of turkey breeding draws on an account previously published in Peter Singer and Jim Mason, *The Ethics of What We Eat* (New York: Random House, 2007).
46. *Poultry Tribune*, January 1974.
47. *Farmer and Stockbreeder*, January 30, 1982; quoted by Harrison, *Animal Machines*, 50.
48. The U.S. figure is an approximation derived from the U.S. Department of Agriculture report *Chickens and Eggs*, February 25, 2022, which shows 624 million egg-type chicks hatched in 2021 (p. 16), coupled with the assumption that half of these chicks will be male. For the EU, the figure is taken from a question asked in the European Parliament by Harald Vilimsky on December 5, 2018: www.europarl.europa.eu/doceo/document/E-8–2018–006133_EN.html.
49. Hannah Thomson, "Fifty million male chicks saved as France bans egg industry from culling," *Connexion* February 8, 2022; "Germany bans male chick culling from 2022, *DW,* dw.com/en/germany-bans-male-chick-culling-from-2022/a-57603148; "Italy bans the killing of male chicks in an effort led by Animal Equality," https://animalequality.org/news/italy-bans-the-killing-of-male-chicks.
50. H. Cheng, "Pain in Chickens and Effects of Beak Trimming," in American College of Poultry Veterinarians, Workshop Proceedings Laying Hen and Pullet Well-being, Management and Auditing, April 18, 2010, (Vancouver, British Columbia), 20, www.ars.usda.gov/research/publications/publication/?seqNo115=253565; see also Taylor Reed, "Beak Trimming: Hot Blade v IRBT," www.slideshare.net/TaylorReed18/beak-trimming, Summer 2016.
51. C. H. Oka, et al., "Performance of Commercial Laying Hen Submitted to Different Debeaking Methods," *Brazilian Journal of Poultry Science* 19 (4): 717–24 (October–December 2017); Philip Glatz and Greg Underwood, "Current methods and techniques of beak trimming laying hens, welfare issues and alternative approaches," *Animal Production Science* 61: 968–89 (2021).
52. See, for example, M. J. Gentle, L. N. Hunter, and D. Waddington, "The onset of pain related behaviours following partial beak amputation in the chicken," *Neuroscience letters* 128 (1): 113–16 (1991).
53. *Report of the Technical Committee to Enquire into the Welfare of Animals Kept*

Under Intensive Livestock Husbandry Systems, Command Paper 2836 (London: Her Majesty's Stationery Office, 1965), paragraph 97.

54. A. Andrade and J. Carson, "The Effect of Age and Methods of Debeaking on Future Performance of White Leghorn Pullets," *Poultry Science* 54 (3): 666–674 (1975); M. Gentle, B. Huges, and R. Hubrecht, "The Effect of Beak Trimming on Food Intake, Feeding Behavior and Body Weight in Adult Hens," *Applied Animal Ethology* 8 (1–2): 147–159 (1982); M. Gentle, "Beak Trimming in Poultry," *World's Poultry Science Journal* 42 (3): 268–275 (1986).
55. J. Breward and M. Gentle, "Neuroma Formation and Abnormal Afferent Nerve Discharges After Partial Beak Amputation (Beak Trimming) in Poultry," *Experienta* 41 (9): 1132–34 (September 1985).
56. N. J. Beausoleil, S. E. Holdsworth, and H. Lehmann, "Avian Nociception and Pain," in *Sturkie's Avian Physiology* (Amsterdam, The Netherlands: Elsevier, 2022), 223–231; see also American Veterinary Medical Association, "Welfare Implications of Beak Trimming," February 7, 2010, www.avma.org/resources-tools/literature-reviews/welfare-implications-beak-trimming.
57. Gentle, "Beak Trimming in Poultry."
58. C. E. Ostrander and R. J. Young, "Effects of Density on Caged Layers," *New York Food and Life Sciences* 3 (3) (1970).
59. *USDA Egg Markets Overview,* July 29, 2022, 4.
60. Marie-Laure Augère-Granier, European Parliamentary Research Service, *The EU Poultry Meat and Egg Sector* PE 644–195, November 2019, 10.
61. *Der Spiegel* 47 (1980): 264; quoted in, *Intensive Egg and Chicken Production* (Huddersfield, UK: Chickens' Lib, 1982).
62. I. Duncan and V. Kite, "Some Investigations into Motivation in the Domestic Fowl," *Applied Animal Behaviour Science* 18 (3–4): 387–388 (1987).
63. *New Scientist* January 30, 1986, 33, reporting on a study by H. Huber, D. Fölsch, and U. Stahli, published in *British Poultry Science* 26 (3): 367 (1985).
64. A. Black and B. Hughes, "Patterns of Comfort Behaviour and Activity in Domestic Fowls: A Comparison Between Cages and Pens," *British Veterinary Journal* 130 (1): 23–33 (1974).
65. The information on dustbathing in this paragraph draws on D. van Liere and S. Bokma, "Short-term Feather Maintenance as a Function of Dustbathing in Laying Hens," *Applied Animal Behaviour Science* 18 (2): 197–204 (1987); H. Simonsen, K. Vestergaard, and P. Willeberg, "Effect of Floor Type Density on the Integument of Egg Layers," *Poultry Science* 59 (10): 2202–06 (1980); and K. Vestergaard, "Dustbathing in the Domestic Fowl—Diurnal Rhythm and Dust Deprivation," *Applied Animal Ethology* 8 (5): 487–95 (1982).
66. Albert Schweitzer Foundation, "Ending the Use of Battery Cages," albertschweitzerfoundation.org/campaigns/ending-use-battery-cages.
67. Eurogroup for Animals, "European commission announces historic commitment to ban cages," Press Release, June 30, 2021, www.eurogroupforanimals

.org/news/european-commission-announces-historic-commitment-ban-cages-farmed-animals.

68. United Egg Producers, "UEP certified conventional cage program: UEP certified guidelines," uepcertified.com/conventional-cage-housing/ (accessed March 20, 2022).
69. B. M. Freeman, "Floor Space Allowance for the Caged Domestic Fowl," *Veterinary Record* 112 (24): 562–63 (June 1983).
70. M. Dawkins, "Do Hens Suffer in Battery Cages? Environmental Preferences and Welfare," *Animal Behaviour* 25 (4): 1034–46 (November 1977). See also M. Dawkins, *Animal Suffering: The Science of Animal Welfare* (London: Chapman and Hall, 1980), chapter 7.
71. Chris Harris, "Finding the Value in Processing Spent Hens," Poultry Site, December 20, 2019, www.thepoultrysite.com/articles/finding-the-value-in-processing-spent-laying-hens.
72. EFSA Panel on Animal Health and Welfare, "Scientific opinion on the killing for purposes other than slaughter: poultry," *EFSA Journal* 17 (11):5850 (2019). The passage quoted is on p.15.
73. Jia-Rui Chong, "Vet in row after hens 'chipped' to death," *Los Angeles Times* November 23, 2003; www.hsus.org/farm_animals/farm_animals_news/missouri_county_files_charges_against_moark.html.
74. United Egg Producers "Animal Husbandry Guidelines for U.S. Egg-Laying Flocks" (2017), uepcertified.com/wp-content/uploads/2021/08/CF-UEP-Guidelines_17–3.pdf; European Commission Guide to good practices for the transport of the Poultry (2018).
75. "Take feed away from spent hens," *Poultry Tribune*, March 1974.
76. Sean Remos, Matthew MacLachlan, and Alex Melton, "Impacts of the 2014–15 Highly Pathogenic Avian Influenza Outbreak in the U.S. Poultry Sector," U.S. Department of Agriculture, *Livestock, Dairy, and Poultry Outlook* (LDPM-282–02) (December 2017).
77. Tom Cullen, "Five million layers snuffed as avian influenza hits," *Storm Lake Times*, March 23, 2022.
78. Gwendolen Reyes-Illg, et al., "The rise of heatstroke as a method of depopulating pigs and poultry: Implications for the US veterinary profession," *Animals* 13: 140 (2023).
79. Marina Bolotnikova, "Amid Bird Flu Outbreak, Meat Producers Seek 'Ventilator Shutdown' For Mass Chicken Killing," *Intercept*, April 14, 2022, theintercept.com/2022/04/14/killing-chickens-bird-flu-vsd/.
80. Marina Bolotnikova, "US farms lobby to use 'cruellest' killing method as bird flu rages," *The Guardian*, November 9, 2022.
81. Our Honor *Weekly Newsletter*, November 30, 2022 and a similar blog-post available at www.ourhonor.org/blognew/avma-denies-petition-of-278-veterinarians-to-reclassify-mass-killing-of-animals-via-heatstroke-as-not-recommended.
82. EFSA Panel on Animal Health and Welfare, "Scientific opinion on the

killing for purposes other than slaughter: poultry," *EFSA Journal* 17 (11):5850 (2019). The passage quoted is on p.25.

83. L214 Media Release, "Millions of birds asphyxiated," April 5, 2022; *"Grippe aviaire. En Vendée, il a dû asphyxier ses 18 000 poulets, puis les enterrer dans un champ,"* ["Avian Flu. In Vendée, he had to asphyxiate his 18,000 chickens, then bury them in a field."], *Ouest France* March 25, 2022.
84. R. Dunbar, "Farming Fit for Animals," *New Scientist* 102: 12–15 (March 1984); D. Wood-Gush, "The Attainment of Humane Housing for Farm Livestock," in M. Fox and L. Mickley, eds., *Advances in Animal Welfare Science* (Dordrecht, The Netherlands: Springer, 1985), 47–55; Gary Landsberg and Sagi Denenberg, "Social Behavior of Swine," *Merck Veterinary Manual*, 2022, www.merckvetmanual.com/behavior/normal-social-behavior-and-behavioral-problems-of-domestic-animals/social-behavior-of-swine.
85. D. Wood-Gush and R. Beilharz, "The Enrichment of a Bare Environment for Animals in Confined Conditions," *Applied Animal Ethology* 10 (3): 209–217 (May 1983); see also R. Dantzer and P. Mormede, "Stress in Farm Animals: A Need for Reevaluation," *Journal of Animal Science* 57 (1): 6–18 (July 1983); and for more recent findings, Marek Špinka, ed., *Advances in Pig Welfare* (Sawston, UK: Woodhead Publishing, 2017), especially Marc Bracke, "Chains as proper enrichment for intensively farmed pigs?," 167–97.
86. Mhairi Sutherland and Cassandra Tucker, "The long and short of it: A review of tail docking in farm animals," *Applied Animal Behaviour Science* 135 (3): 179–917 (December 2011).
87. D. Fraser, "The role of behaviour in swine production: a review of research," *Applied Animal Ethology* 11 (4): 317–339 (1984).
88. D. Fraser, "Attraction to blood as a factor in tail-biting by pigs," *Applied Animal Behaviour Science* 17 (1–2): 61–68 (1987).
89. M. Larsen, H. Andersen, and L. Petersen, "Which is the most preventive measure against tail docking in finisher pigs: tail docking, straw provision or lowered stocking density?," *Animal* 12 (6): 1260–67 (June 2018).
90. L214, "*Historique: une infraction routinière enfin condamnée*," savoir-animal.fr/historique-une-infraction-routiniere-enfin-condamnee/. See also European Court of Auditors, "Animal welfare in the EU: closing the gap between ambitious goals and practical implementation," *Special Report* No. 31 (2018).
91. The figure of 93.5 percent is based on data from Table 20 of the U.S. Department of Agriculture's 2017 Census of Agriculture, which shows a total of 235.3 million pigs sold that year. Of that total, 220.1 million were sold from farms with more than five thousand pigs. The census is available at www.nass.usda.gov/Publications/AgCensus/2017/Full_Report/volume_1,_chapter_1_US/usv1.pdf.
92. Tyson Food Facts page, ir.tyson.com/about-tyson/facts/default.aspx, accessed August 11, 2022.

93. "Life cycle of a market pig," porkcheckoff.org/pork-branding/facts-statistics /life-cycle-of-a-market-pig/, accessed 26 March 2022.
94. *Hog Farm Management*, December 1975, 16.
95. Bob Fase, quoted in Orville Schell, *Modern Meat* (New York: Random House, 1984), p.62.
96. Lauren Kendrick, *Ammonia Emissions from Industrial Hog Farming* (Santa Monica, CA: RAND Corporation, 2018).
97. *National Hog Farmer*, March 1978, 27.
98. U. S. Department of Agriculture, *Fact Sheet, Swine Management*, AFS-3–12 (Washington, D.C.: Office of Governmental and Public Affairs, 1981), 1.
99. K. H. Kim, et al., "Effects of Gestational Housing on Reproductive Performance and Behavior of Sows with Different Backfat Thickness," *Asian-Australasian Journal of Animal Science* 29: 142–48 (2016); Temple Grandin is quoted in Mark Essig, "Pig farming doesn't have to be this cruel," *New York Times*, October 10, 2022.
100. Commission of the European Communities. Scientific Veterinary Committee, *The Welfare of Intensively Kept Pigs: Report of the Scientific Veterinary Committee, Adopted 30 September 1997.*
101. U.S. Department of Agriculture, Economic Research Service, "Hog welfare laws cover 9 states and 3 percent of the national herd in 2022," last updated March 16, 2022.
102. U.S. Department of Agriculture, *Fact Sheet: Swine Housing*, AFS-3–8–9 (Washington D.C.: Office of Governmental and Public Affairs, 1981), 4.
103. A. Lawrence, M. Appleby, and H. MacLeod, "Measuring hunger in the pig using operant conditioning: The effect of food restriction," *Animal Science* 47 (1): 131–37(1988).
104. Commission of the European Communities. Scientific Veterinary Committee, *The Welfare of Intensively Kept Pigs.*
105. Eric Schlosser, reviewing Corban Addison's *Wastelands* in the *New York Times,* June 7, 2022.
106. U.S. Department of Agriculture, Food, Safety and Inspection Service, "Veal from Farm to Table," last updated August 2013, www.fsis.usda.gov /food-safety/safe-food-handling-and-preparation/meat/veal-farm-table.
107. See the April 1973 and November 1973 issues of *Stall Street Journal*, the newsletter of Provimi Inc., at the time the largest U.S. producer of white veal.
108. "American Veal Association Confirms Mission Accomplished," January 2018, www.americanveal.com/industry-updates/2018/1/22/american-veal -association-confirms-mission-accomplished.
109. See the video "What do calves eat?"at www.americanveal.com/veal-videos.
110. Email from James Reynolds to Sophie Kevany, April 19, 2022.
111. M. Shahbandeh, "Per Capita Consumption of Veal in the United States," Statista, March 2, 2022, https://www.statista.com/statistics/183541/per -capita-consumption-of-veal-in-the-us; U.S. Department of Agriculture,

"Ask USDA: How much veal is consumed in the United States," https://ask.usda.gov/s/article/How-much-veal-is-consumed-in-the-US; "Veal: Production and Consumption in Europe," https://www.vealthebook.com/process/production-and-consumption-in-europe. (Both sites accessed October 7, 2022.)

112. Compassion in World Farming, "Standard Intensive Milk Production," www.compassioninfoodbusiness.com/awards/good-dairy-award/standard-intensive-milk-production/. For 2014 statistics, the most recent available, see U.S. Department of Agriculture, *Dairy 2014: Dairy Cattle Management Practices in the United States, 2014*, www.aphis.usda.gov/animal_health/nahms/dairy/downloads/dairy14/Dairy14_dr_PartI_1.pdf.
113. See, for example. Ontario Ministry of Agriculture, Food and Rural Affairs, "Lighting for More Milk," Agdex 717, reviewed January 2019, www.omafra.gov.on.ca/english/engineer/facts/06–053.htm.
114. James MacDonald, "Scale economies provide advantages to large dairy farms," U.S. Department of Agriculture, Economic Research Service, August 3, 2020, www.ers.usda.gov/amber-waves/2020/august/scale-economies-provide-advantages-to-large-dairy-farms/.
115. Lyndal Reading, "How Now dairy: Taking an animal welfare approach on milking, *Weekly Times*, November 6, 2017; see also https://hownowdairy.com.au.
116. For Ahimsa, see www.ahimsamilk.org/. Anja Hradetsky's farm is featured in "Who we eat: The Status Quo," directed by Jannis Funk and Jakob Schmidt and made by Arte in 2021.
117. *Encyclopedia Britannica*, www.britannica.com/animal/cow.
118. *Peoria Journal Star*, June 5, 1988.
119. U.S. Department of Agriculture, Economic Research Service, "Cattle & Beef: Sector at a Glance" (last updated November 2021), www.ers.usda.gov/topics/animal-products/cattle-beef/sector-at-a-glance.
120. U.S. Department of Agriculture, Economic Research Service, "Cattle & Beef: Sector at a Glance"; Temple Grandin, "Evaluation of the welfare of cattle housed in outdoor feedlot pens," *Veterinary and Animal Science* 1–2: 23–28 (December 2016).
121. Beef Cattle Research Council, "Acidosis," www.beefresearch.ca/research-topic.cfm/acidosis-63 (accessed August 26, 2022); L.C. Eastwood, et al., "National Beef Quality Audit-2016: Transportation, mobility, and harvest-floor assessments of targeted characteristics that affect quality and value of cattle, carcasses, and by-products," *Translational Animal Science* 1 (2): 229–38 (April 2017).
122. Lily Edwards-Callaway, et al., "Impacts of shade on cattle well-being in the beef supply chain," *Journal of Animal Science* 99 (2): 1–21 (February 2021).
123. F. M. Mitlöhner, et al., "Effects of shade on heat-stressed heifers housed under feedlot conditions," Burnett Center Internet Progress Report, no. 11, February 2001; www.depts.ttu.edu/afs/burnett_center/progress_reports

/bc11.pdf; see also F. M. Mitlöhner, et al., "Shade effects on performance, carcass traits, physiology, and behavior of heat-stressed feedlot heifers," *Journal of Animal Science,* 80 (8): 2043–50 (August 2002).

124. Elisha Fieldstadt and Reuters, "At least 2000 cattle dead in Kansas heat, adding pain to beleaguered industry," NBC News, June 18, 2022, www.nbcnews.com/news/weather/least-2000-cattle-dead-kansas-heat-adding-pain-beleaguered-industry-rcna33877.
125. Temple Grandin, "Evaluation of the welfare of cattle housed in outdoor feedlot pens."
126. European Food Safety Authority, "General approach to fish welfare and to the concept of sentience in fish, Scientific Opinion of the Panel on Animal Health and Welfare," adopted January 29, 2009, *EFSA Journal* 954: 1–27 (2009).
127. Alison Mood et al, "Estimating global numbers of farmed fishes killed for food annually from 1990 to 2019," *Animal Welfare,* 32 e12 (2023) 1-16; see also B. Franks, C. Ewell, and J. Jacquet, "Animal welfare risks of global aquaculture," *Science Advances* 7: eabg0677 (2021).
128. Annie Rueter, et al., "The Fish You Don't Know You Eat," produced by NBC News and the Global Reporting Program at the University of British Columbia Graduate School of Journalism, globalreportingprogram.org/fishmeal/.
129. Lao Minyi, "Fish farms are transformed into laboratories: fish farmers using microscopes to find parasites and prescribe the right medicine to increase the survival rate to 50%," *Hong Kong News* July 20, 2019, www.hk01.com/sns/article/352888.
130. See this spreadsheet compiled by Persis Eskander for the Open Philanthropy Project, based on 2016 figures from the United Nations Food and Agriculture Organization: docs.google.com/spreadsheets/d/12pA0UxIbRDcfY5g25XZ7na4duhj6411l-1-_3tRH48k/edit#gid=1419062790.
131. J. Lines & J. Spence, "Safeguarding the welfare of farmed fish at harvest," *Fish Physiology and Biochemistry* 38 (1): 153–62 (February 2012). I owe this reference to the Fish Welfare Initiative, *Fish Welfare Improvements in Aquaculture,* December 2020, 10.13140/RG.2.2.17712.58889.
132. Daniela Waldhorn and Elisa Autric, "Shrimp Production: Understanding the Scope of the Problem," a 2023 report for Rethink Priorities, https://rethinkpriorities.org/shrimp-production; see also Becca Franks, Christopher Ewell, and Jennifer Jacquet, "Animal welfare risks of global aquaculture," *Science Advances* 7: eabg0677(2021).
133. For further discussion see Almaya Albalat, et al., "Welfare in Farmed Decapod Crustaceans, With Particular Reference to *Penaeus vannamei*" *Frontiers in Marine Science* 9: 886024 (2022). See also the sources in note 17 to chapter 1.
134. Jennifer Jacquet, Becca Franks, and Peter Godfrey-Smith, "The octopus mind and the argument against farming it," *Animal Sentience* 26: 19 (2019).

135. Paulo Steagall, et al., "Pain Management in Farm Animals: Focus on Cattle, Sheep and Pigs," *Animals* 11 (6): 1483 (May 2021).
136. Calla Wahlquist, "'Horrific' footage of live cattle having horns removed in Australia sparks outrage," *Guardian,* December 5, 2019.
137. Animal Outlook, "Animal Transport: Torture Hidden in Plain Sight," animaloutlook.org/investigations/#transport. The statement about the lack of prosecutions is based on information from Piper Hoffman, senior director of legal advocacy for Animal Outlook, in an email to Sophie Kevany, February 7, 2022. See also these Animal Welfare Institute documents: awionline.org/sites/default/files/uploads/documents/AWI-Request-to-Enforce-28-Hour-Law.pdf; awionline.org/sites/default/files/uploads/documents/21LegalProtectionsTransport.pdf.
138. Barbara Padalino, et al., "Transport certifications of cattle moved from France to Southern Italy and Greece: do they comply with Reg. EC 1/2005?," *Italian Journal of Animal Science* 20 (1): 1870–81 (October 2021).
139. Karen Schwartzkopf-Genswein, et al., "Symposium Paper: Transportation issues affecting cattle well-being and considerations for the future," *Professional Animal Scientist* 32 (6): 707–16 (December 2016).
140. Timothy Pachirat, *Every Twelve Seconds: Industrialized Slaughter and the Politics of Sight,* (New Haven, CT: Yale University Press, 2011), 145.
141. Pachirat, *Every Twelve Seconds,* 60.
142. Pachirat, *Every Twelve Seconds,* 153–56.
143. Pachirat, *Every Twelve Seconds,* 85–86.
144. Animal Equality UK, "Our Investigations into Slaughterhouses, https://animalequality.org.uk/our- investigations-into-slaughterhouses.
145. L214, *"L'enfer des veaux à l'abattoir Sobeval,"* www.l214.com/enquetes/2020/abattoir-veaux-sobeval/.
146. Alison Mood, *Worse things happen at sea: The welfare of wild-caught fish,* 2010, www.fishcount.org.uk/published/standard/fishcountsummaryrptSR.pdf.
147. J. Lines and J. Spence, "Safeguarding the welfare of farmed fish at harvest," *Fish Physiology and Biochemistry* 38: 153–62 (February 2012).
148. European Food Safety Authority, "Species-specific welfare aspects of the main systems of stunning and killing of farmed eels, *Anguilla Anguilla;* Scientific opinion of the panel on animal health and welfare," *EFSA Journal* 1014: 1–42 (2009).
149. See, for the Netherlands, *Regulation of the Minister of Agriculture, Nature and Food Quality of 15 May 2018, no. WJZ/17127055, amending the Regulation on animal keepers in connection with the stunning of eels,* https://zoek.officielebekendmakingen.nl/stcrt-2018–25060.html; New Zealand Government, *Code of Welfare: Commerical Slaughter,* p.28, www.mpi.govt.nz/dmsdocument/46018-Code-of-Welfare-Commercial-slaughter. In Germany, this website for anglers says that various painful methods of killing eels, including the use of salt, are "plainly and simply not permitted": www.netzwerk-angeln.de/angeln/fischverwertung/194-aal-toeten-und-ausnehmen.html.

150. See the video at www.youtube.com/watch?v=JdF12LQ4wGA.
151. See the video at www.youtube.com/watch?v=hbW88hGyDEU; translation is provided on the video.

CHAPTER 4: LIVING WITHOUT SPECIESISM . . .

1. See Will MacAskill, *Doing Good Better: How Effective Altruism Can Help You Help Others, Do Work that Matters, and Make Smarter Choices about Giving Back* (New York: Penguin Random House, 2016); and Peter Singer, *The Most Good You Can Do* (New Haven, CT: Yale University Press, 2015).
2. For advice on careers that do the most good, see 80000hours.org.
3. Our World in Data, "Global Meat Consumption, 1961–2050," https://ourworldindata.org/grapher/global-meat-projections-to-2050?time=1961.2050.
4. The information on meat consumption is based on figures for meat availability from the database of the Organization for Economic Cooperation and the Food and Agriculture Organization of the United Nations.
5. People for the Ethical Treatment of Animals, "Air France Commits to Banning Transport of Monkeys to Laboratories," June 30, 2022, www.peta.org/about-peta/victories/.
6. Maria Salazar, "The Effects of Diet Choices," Animal Charity Evaluators, February 2021, based on 2018 figures. Available at animalcharityevaluators.org/research/reports/dietary-impacts/effects-of-diet-choices.
7. Such arguments are discussed by Julia Driver, Mark Budolfson, and Clayton Littlejohn in their essays in Ben Bramble and Bob Fischer, eds., *The Moral Complexities of Eating Meat* (Oxford University Press, 2015).
8. Bailey Norwood and Jayson Lusk, *Compassion, by the Pound: The Economics of Farm Animal Welfare* (Oxford University Press, 2011), chapter 8.
9. Hannah Ritche and Max Roser, "Soy," Our World in Data, ourworldindata.org/soy, accessed October 9 2022.
10. A. Shepon, et al., *Environmental Research Letters* 11 (10): article 105002 (2016); A. Shepon, et al., "The opportunity costs of animal based foods exceeds all food losses," *Proceedings of the National Academy of Sciences of the U.S.A.* 115 (15): 3804–09 (April 10, 2018).
11. P. Alexander, et al., "Human appropriation of land for food: The role of diet," *Global Environmental Change* 41: 88–98 (November 2016). For the percentage of global habitable land area needed for agriculture if the total world population were to adopt the average diet of a given country, see Hannah Ritchie, "How much of the world's land would we need in order to feed the global population with the average diet of a given country?" Our World in Data, October 3, 2017, ourworldindata.org/agricultural-land-by-global-diets. See also see Michael Grunwald, "No one wants to say 'Put down that burger,' but we really should," *New York Times*, December 15, 2022
12. Food and Agriculture Organization, *The State of World Fisheries and Aquaculture 2022. Towards Blue Transformation*, Rome, FAO, 2022, https://doi.org/10.4060/cc0461en.

13. Global Initiative Against Transnational Organized Crime, "Illicit Migration to Europe: Consequences of illegal fishing and overfishing in West Africa," May 8, 2015; globalinitiative.net/analysis/illicit-migration-to-europe-consequences-of-illegal-fishing-and-overfishing-in-west-africa/, accessed 14 August 2022.
14. Munk Debate, Animal Rights, guests Peter Singer and Joel Salatin, January 25, 2022, munkdebates.com/podcast/animal-rights.
15. Jacy Reese Anthis, "U.S. Factory Farming Estimates," Sentience Institute 2019, https://www.sentienceinstitute.org/us-factory-farming-estimates.
16. Information from Compassion in World Farming, available at www.ciwf.org.uk/farm-animals/sheep/.
17. Stone Barns Center for Food and Agriculture, "Back to grass: the market potential for U.S. grassfed beef," 2017, www.stonebarnscenter.org/blog/future-grassfed-beef-green; Meat and Livestock Australia, "Grainfed cattle make up 50% of beef production," June 10, 2021, https://www.mla.com.au/prices-markets/market-news/2021/grainfed-cattle-make-up-50-of-beef-production/.
18. See Leslie Stephen, *Social Rights and Duties* (London, 1896), and quoted by Henry Salt, "The Logic of the Larder," which appeared in Salt's *The Humanities of Diet* (Manchester: The Vegetarian Society, 1914), 34–38, and has been reprinted in Tom Regan and Peter Singer, eds., *Animal Rights and Human Obligations* (Englewood Cliffs, NJ: Prentice-Hall, 1976).
19. Roger Scruton, "The Conscientious Carnivore," in S. Sapontzis, ed., *Food for Thought* (Amherst, MA: Prometheus, 2004), 81–91; see also Roger Scruton, *Animal Rights and Wrongs* (London: Continuum, 2006); Michael Pollan, *The Omnivore's Dilemma: A Natural History of Four Meals* (New York Penguin Books, 2016); and G. Schedler, "Does ethical meat eating maximize utility?," *Social Theory and Practice* 31 (4): 499–511 (October 2005).
20. For attempts to do this, see S. Sapontzis, *Morals, Reason and Animals* (Philadelphia: Temple University Press, 1987), 193–94; David Benatar, *Better Never to Have Been: The Harm of Coming into Existence* (Oxford University Press, 2009), chapter 2; and Melinda Roberts, "An Asymmetry in the Ethics of Procreation," *Philosophy Compass* 6 (11): 765–76 (November 2011). For discussion, see Peter Singer, *Practical Ethics*, 3rd ed. (Cambridge University Press, 2011, chapters 4 and 5.
21. For more on the ideas discussed in this paragraph, see Henry Sidgwick, *The Methods of Ethics*, (London: Macmillan, 1907), book IV, chapter 1; Derek Parfit, *Reasons and Persons* (Oxford: Clarendon Press, 1984), Part IV; Gustaf Arrhenius, Jesper Ryberg, and Torbjörn Tännsjö, "The Repugnant Conclusion," *The Stanford Encyclopedia of Philosophy*, Summer 2022 ed., Edward N. Zalta ed., plato.stanford.edu/archives/sum2022/entries/repugnant-conclusion; Jeff McMahan, "Eating Animals the Nice Way, *Daedalus* Winter 2008, 1–11; Tatjana Visak, *Killing Happy Animals* (Houndsmill, Bas-

ingstoke: Palgrave Macmillan, 2013); and Andy Lamey, *Duty and the Beast*, (Cambridge: Cambridge University Press, 2019), especially chapters 5 and 7.

22. I owe this last point to Adam Lerner, "The Procreation Asymmetry Asymmetry" [sic], *Philosophical Studies*, forthcoming.

23. Oliver Goldsmith, *The Citizen of the World*, in *Collected Works*, volume 2, A. Friedman, ed., (Oxford: Clarendon Press, 1966), 60. Apparently Goldsmith himself fell into this category, however, since according to Howard Williams in *The Ethics of Diet* (abridged edition, Manchester and London, 1907, 149), Goldsmith's sensibility was stronger than his self-control.

24. Intergovernmental Panel on Climate Change, *Climate Change 2022: Mitigation of Climate Change*, 2022, Technical Summary, TS-89, www.ipcc.ch/report/ar6/wg3/.

25. Laura Wellesley and Antony Froggatt, *Changing Climate, Changing Diets: Pathways to Lower Meat Consumption* (London: Chatham House, 2015). The summary is quoted from www.chathamhouse.org/2015/11/changing-climate-changing-diets-pathways-lower-meat-consumption.

26. See Hannah Ritchie, "You want to reduce the carbon footprint of your food? Focus on what you eat, not whether your food is local," ourworldindata.org/food-choice-vs-eating-local, drawing on J. Poore and T. Nemecek, "Reducing food's environmental impacts through producers and consumers," *Science* 360 (6392): 987–92, and V. Sandström, et al., "The role of trade in the greenhouse gas footprints of EU diets," *Global Food Security* 19: 48–55 (December 2018).

27. Hannah Ritchie and Max Roser, "Environmental Impacts of Food Production," Our World in Data, ourworldindata.org/environmental-impacts-of-food#carbon-footprint-of-food-products.

28. For the study of the carbon opportunity costs of animal agriculture, see Mathew Hayek, et al., "The carbon opportunity cost of animal-sourced food production on land," *Nature Sustainability* 4: 21–24 (2021); for the other study referred to in this paragraph, including the quotation, see Michael Eisen and Patrick Brown, "Rapid global phaseout of animal agriculture has the potential to stabilize greenhouse gas levels for 30 years and offset 68 percent of CO2 emissions this century," *PLOS Climate* 1 (2): e0000010 (2022).

29. Poore is quoted in Damian Carrington, "Avoiding meat and dairy is 'single biggest way' to reduce your impact on Earth," *The Guardian*, June 1, 2018. The study Poore led is in J. Poore and T. Nemecek, "Reducing food's environmental impact through producers and consumers," *Science* 360 (6392): 987–92 (June 2018).

30. See The Fish Site, "The case against eyestalk ablation in shrimp aquaculture," September 22, 2020, thefishsite.com/articles/the-case-against-eyestalk-ablation-in-shrimp-aquaculture; on shrimp welfare more generally, see www.shrimpwelfareproject.org/; see also G. Diarte-Plata, et al., "Eyestalk ablation procedures to minimize pain in the freshwater prawn

Macrobrachium americanum," *Applied Animal Behaviour Science* 140 (3–4): 172–78 (2012).

31. Katherine Martinko, "Why it's a good idea to stop eating shrimp," *Treehugger*, May 5, 2020, https://www.treehugger.com/shrimp-may-be-small-their-environmental-impact-devastating-4858308.
32. E. S. Nielsen and L. A. Mound, "Global diversity of insects: the problems of estimating numbers," in: P. H. Raven and Tania Williams, eds., *Nature and Human Society: The Quest for a Sustainable World* (Washington, D.C.: National Academy Press, 2000), 213–22.
33. Abraham Rowe, "Insects raised for food and feed—global scale, practices, and policy," *Rethink Priorities*, June 29, 2020, rethinkpriorities.org/publications/insects-raised-for-food-and-feed.
34. A. van Huis, "Welfare of farmed insects," *Journal of Insects for Food and Feed* 5 (3): 159–62 (2019).
35. Berthold Hedwig and Stefan Schöneich, "Neural circuit in the cricket brain detects the rhythm of the right mating call," University of Cambridge, Research, September 11, 2015, www.cam.ac.uk/research/news/neural-circuit-in-the-cricket-brain-detects-the-rhythm-of-the-right-mating-call#:~:text=Using%20tiny%20electrodes%2C%20scientists%20from,up%20to%20a%20million%20neurons.
36. Peter Godfrey-Smith, "Somewhere between a shrimp and an oyster," *Metazoan* 61 (April 2018), metazoan.net/61-somewhere-between/.
37. Chesapeake Bay Program, "Oysters," www.chesapeakebay.net/issues/oysters; see also Jennifer Jacquet, et al., "Seafood in the future: Bivalves are better," *Solutions*, January 11, 2017, thesolutionsjournal.com/2017/01/11/seafood-future-bivalves-better.
38. Some of the recent articles and blogs that have influenced my thinking on this topic, in addition to those already cited, are: Christopher Cox, "Consider the oyster," *Slate* 7 (April 2010); Diana Fleischman, "The ethical case for eating oysters and mussels," published 2013 and reposted 2020, Pt 1, dianaverse.com/2020/04/07/bivalveganpart1/, Pt 2 dianaverse.com/2020/04/07/bivalveganpart2; David Cascio, "On the consumption of bivalves," *Animalist* (January 20, 2017), theanimalist.medium.com/on-the-consumption-of-bivalves-bdde8db6d4ba; and Brian Tomasik, "Can bivalves suffer?" (February 2017, updated 2019), reducing-suffering.org/can-bivalves-suffer. For the percentages of bivalves farmed, see Food and Agriculture Organization of the United Nations, *Fishery and Agriculture Statistics Yearbook 2019*, FAO, 2021. I thank Alessandra Roncarati, co-author of "Global Production of Marine Bivalves. Trends and Challenges,"in Aad Smaal, et al., *Goods and Services of Marine Bivalves* (Dordrecht, The Netherlands: Springer, 2018), for this information.
39. "Nick Kyrgios on why he is vegan," BBC Sport, June 30, 2022, www.facebook.com/watch/?v=1500784720364306.

40. M. J. Orlich, et al., "Vegetarian dietary patterns and mortality in Adventist Health Study 2," *JAMA Internal Medicine* 173(13): 1230–38.
41. A. Satija, et al., "Plant-Based Dietary Patterns and Incidence of Type 2 Diabetes in U.S. Men and Women: Results from Three Prospective Cohort Studies," *PLoS Medicine* 13 (6): e1002039 (June 2016); A. Satija, et al., "Healthful and Unhealthful Plant-Based Diets and the Risk of Coronary Heart Disease in U.S. Adults," *Journal of the American College of Cardiology* 70 (4): 411–22 (July 2017).
42. Walter Willett, et al., "Food in the Anthropocene: the EAT–*Lancet* Commission on healthy diets from sustainable food systems," *Lancet* 393 (10170): 447–92 (February 2019).
43. The EAT-Lancet Commission, *Food, Planet, Health: Summary Report of the EAT-Lancet Commission,* available at eatforum.org/eat-lancet-commission/.

CHAPTER 5: MAN'S DOMINION . . .

1. The Buddhist perspective is clearly more compassionate to animals than the Western tradition has been, even though relatively few of those who consider themselves Buddhist live in accordance with the Buddha's teachings in this respect. A dialogue I had with Shih Chao-Hwei that discusses this issue will soon be published as Peter Singer and Shih Chao-Hwei, *The Buddhist and the Ethicist* (Boulder, CO: Shambala Publications, 2023). For an account of the many Islamic texts encouraging for kind treatment of animals and providing support for the idea that, as the Quran says, animals are "communities like yourselves," see Al-Hafiz B.A. Masri, *Animals in Islam*, Nadeem Haque, ed. (Woodstock and Brooklyn, NY: Lantern Publishing and Media, 2022).
2. Gen. 1:26.
3. Gen. 9:1–3.
4. Aristotle, *Politics*, Everyman's Library (London: J. M. Dent & Sons, 1956), 10.
5. Aristotle, *Politics*, 16.
6. W. E. H. Lecky, *History of European Morals from Augustus to Charlemagne,* volume I (London: Longmans, 1869), 280–82.
7. Mark 5:1–13.
8. Cor. 9:9–10.
9. Saint Augustine, *The Catholic and Manichaean Ways of Life*, D. A. Gallagher and I. J. Gallagher, trans. (Boston: The Catholic University Press, 1966), 102. I owe this reference to John Passmore, *Man's Responsibility for Nature* (New York: Scribner's, 1974), 11.
10. W. E. H. Lecky, *History of European Morals*, volume I, 244; for Plutarch, see especially the essay "On the Eating of Flesh" in his *Moral Essays*; for Apuleius, I have produced an abridged edition to focus particularly on the tale of the donkey: Apuleius, *The Golden Ass*, Peter Singer, ed., Ellen Finkelpearl, trans. (New York: Norton, 2021).

11. For Basil, see John Passmore, "The Treatment of Animals," *Journal of the History of Ideas* 36 (2): 198 (1975); for Chrysostom, Andrew Linzey, *Animal Rights: A Christian Assessment of Man's Treatment of Animals* (London: SCM Press, 1976), 103; and for Saint Isaac the Syrian, A. M. Allchin, *The World Is a Wedding: Explorations in Christian Spirituality* (London: Darton, Longman and Todd, 1978), 85. I owe these references to R. Attfield, "Western Traditions and Environmental Ethics," in R. Elliot and A. Gare, eds., *Environmental Philosophy* (St. Lucia: University of Queensland Press, 1983), 201–30. For further discussion see Attfield's *The Ethics of Environmental Concern* (Oxford: Blackwell, 1982); K. Thomas, *Man and the Natural World: Changing Attitudes in England 1500–1800* (London: Allen Lane, 1983), 152–53; and R. Ryder, *Animal Revolution: Changing Attitudes Towards Speciesism* (Oxford: Blackwell, 1989), 34–35.
12. See *St. Francis of Assisi, His Life and Writings as Recorded by His Contemporaries*, L. Sherley-Price, trans. (London: Mowbray, 1959), especially p. 145; Wikipedia, "Rule of St Francis," en.wikipedia.org/wiki/Rule_of_Saint_Francis, accessed July 7, 2022.
13. *Summa theologica* II, II, Q64, art. 1.
14. *Summa theologica* II, II, Q159, art. 2.
15. *Summa theologica* I, II, Q72, art. 4.
16. *Summa theologica* II, II, Q25, art. 3.
17. *Summa theologica* II, I, Q102, art. 6; see also *Summa contra gentiles* III, II, 112 for a similar view.
18. The first quote is from Giovanni Pico della Mirandola, *Oration on the Dignity of Man* (1486) and the second is from Marsilio Ficino, *Theologia platonica* (1482), see especially III, 2 and XVI, 3; see also Giannozzo Manetti, *On the Dignity and Excellence of Man* (1453).
19. E. McCurdy, *The Mind of Leonardo da Vinci* (London: Cape, 1932), 78.
20. Michel de Montaigne, *An Apology for Raymond Sebond*, M. A. Screech, trans. (London: Penguin, 1987).
21. René Descartes, *Discourse on Method*, volume 5; see also his letter to Henry More February 5, 1649. I have given the standard reading of Descartes, the way in which his position was understood at the time and has been understood by most of his readers up to and including the present; but this standard reading has been contested. For further details, see John Cottingham, "'A Brute to the Brutes?': Descartes' Treatment of Animals," *Philosophy* 53 (206): 551–59 (1978).
22. *Hansard's Parliamentary History*, April 18, 1800.
23. John Passmore describes the question "Why do animals suffer?" as "for centuries the problem of problems. It engendered fantastically elaborate solutions. Malebranche [a contemporary of Descartes] is quite explicit that for purely theological reasons it is necessary to deny that animals can suffer, since all suffering is the result of Adam's sin and the animals do not descend from Adam." See Passmore, *Man's Responsibility for Nature*, 114n.

24. René Descartes, letter to Henry More, February 5, 1649.
25. Nicholas Fontaine, *Memoires pour servir A l'histoire de Port-Royal*, volume 2 (Cologne, 1738), 52–53, quoted in L. Rosenfield, *From Beast-Machine to Man-Machine: The Theme of Animal Soul in French Letters from Descartes to La Mettrie* (New York: Oxford University Press, 1940).
26. David Hume, An *Enquiry Concerning the Principles of Morals* (1751), chapter 3.
27. Voltaire, "Bêtes," *Dictionnaire Philosophique* (1764).
28. *The Guardian* May 21, 1713.
29. Voltaire, *Elements of the Philosophy of Newton*, volume 5; see also *Essay on the Morals and Spirit of Nations.*
30. Jean-Jacques Rousseau, *Emile*, Everyman's Library (London: J. M. Dent & Sons, 1957), 118–20.
31. Immanuel Kant, *Lectures on Ethics*, L. Infield, trans. (New York: Harper, 1963), 239–40.
32. *Hansard's Parliamentary History*, April 18, 1800.
33. E. S. Turner, *All Heaven in a Rage* (London: Michael Joseph, 1964), 127. Other details in this section come from chapters 9 and 10 of this book.
34. It has been claimed that the first legislation protecting animals from cruelty was enacted by the Massachusetts Bay Colony in 1641. Section 92 of "The Body of Liberties," printed in that year, reads: "No man shall exercise any Tirranny or Crueltie towards any bruite Creature which are usually kept for man's use"; and the following section requires a rest period for animals being driven. This is a remarkably advanced document; one could quibble over whether it was technically a "law," but certainly Nathaniel Ward, compiler of the "Body of Liberties," deserves to be remembered along with Richard Martin as a legislative pioneer. For a fuller account, see Emily Leavitt, *Animals and Their Legal Rights* (Washington, D.C.: Animal Welfare Institute, 1970.
35. Quoted in Turner, *All Heaven in a Rage*. For an exploration of the implications of this remark that adds considerably to the discussion here, see James Rachels, *Created from Animals: The Moral Implications of Darwinism* (Oxford University Press, 1990).
36. Charles Darwin, *The Descent of Man* (London, 1871), 1.
37. Darwin, *Descent of Man*, chapters 3 and 4. The words quoted are from p. 193.
38. William Paley, *The Principles of Moral and Political Philosophy*, volume 2 (1785; reprint, Cambridge University Press, 2013), chapter 11.
39. See Francis Wayland, *Elements of Moral Science* (Cambridge, MA: Harvard University Press, 1963), 364.
40. Quoted by S. Godlovitch, "Utilities," in Godlovitch and Harris, eds., *Animals, Men and Morals.* (New York: Taplinger, 1972)
41. Benjamin Franklin, *Autobiography* (New York: Modern Library, 1950), 41.
42. Quoted in H. S. Salt, *Animals' Rights* (1892), 15.

43. Jules Michelet, *La bible de l'humanité* (1864), quoted in H. Williams, *The Ethics of Diet*, abridged edition, (Manchester and London, 1907), 214.
44. Arthur Schopenhauer, *On the Basis of Morality*, E. F. J. Payne, trans. (Indianapolis: Bobbs-Merrill, Library of Liberal Arts, 1965), 182; see also Arthur Schopenhauer, *Parerga und Paralipomena*, volume 2, chapter 15.
45. Turner, *All Heaven in a Rage*, 143.
46. Turner, *All Heaven in a Rage*, 205.
47. T. H. Huxley, *Man's Place in Nature* (Ann Arbor: University of Michigan Press, 1959), chapter 2.
48. Turner, *All Heaven in a Rage*, 163.
49. V. J. Bourke, *Ethics* (New York: Macmillan, 1951), 352.
50. John Paul II, *Sollicitudo rei socialis* (Homebush, NSW: St. Paul Publications, 1988), 34: 3–74.
51. Quoted in Kurt Remele, "A Strange Kind of Kindness—On Catholicism's Moral Ambiguity Toward Animals," in Andrew Linzey and Clair Linzey, eds., *The Routledge Handbook of Religion and Animal Ethics* (Oxfordshire: Routledge, 2018).
52. Rick Gladstone, "Dogs in Heaven? Pope Francis Leaves Pearly Gates Open," *New York Times*, December 11, 2014.
53. For the views of the philosophers mentioned, see Tom Regan, *The Case for Animal Rights* (Berkeley and Los Angeles: University of California Press, 1983); Carol Adams, *The Sexual Politics of Meat* (New York: Continuum, 1990); Mark Rowlands, *Animal Rights* (New York: St Martin's Press, 1998); Lori Gruen, *Ethics and Animals* (Cambridge: Cambridge University Press, 2011); Christine Korsgaard, *Fellow Creatures* (Oxford: Oxford University Press, 2018); Martha Nussbaum, *Justice for Animals* (New York: Simon and Schuster, 2022); and Alice Crary and Lori Gruen, *Animal Crisis* (Cambridge: Polity Press, 2022).
54. On current views of Chinese philosophers on eating meat, see Tiantian Hou, Xiaojun Ding, and Feng Yu, "The moral behavior of ethics professors: A replication-extension in Chinese mainland," *Philosophical Psychology*, DOI: 10.1080/09515089.2022.2084057 (2022). For the convergence possible between the Buddhist and secular utilitarian traditions, see Peter Singer and Shih Chao-Hwei, *The Buddhist and the Ethicist* (Boulder, CO: Shambhala, 2023).

CHAPTER 6: SPECIESISM TODAY . . .

1. Amy Pixton, *Hello Farm!* (New York: Workman's Publishing, 2018), was ranking No.1 in Children's Farm Life Books on Amazon.com on August 3, 2022. For further examples, see Axel Scheffler, *On the Farm* (London: Campbell Books, 2018), which was the No. 1 bestseller in activity books for children on Amazon.co.uk July 28, 2022: and Thea Feldman, *Discovery: Moo on the Farm* (San Diego: Silver Dolphin Books, 2019).
2. Lawrence Kohlberg, "From Is to Ought," in T. Mischel, ed. *Cognitive Devel-*

opment and Epistemology (New York: Academic Press, 1971), 191–92. Children often see humans and animals as closer in moral status than adults do. See M. Wilks, et al., "Children prioritize humans over animals less than adults do," *Psychological Science* 32 (1): 27–38 (2021).

3. Rick Dewsbury, "Tesco forced to withdraw sausage adverts as pigs 'not as happy' as they claimed," *Daily Mail*, September 14, 2021.
4. Turner, *All Heaven in a Rage*, 129.
5. Turner, *All Heaven in a Rage*, 83.
6. For the historical details in this paragraph, see Turner, *All Heaven in a Rage*, 83, 129; and Gerald Carson, *Cornflake Crusade* (New York Rinehart, 1957), 19, 53–62.
7. Turner, *All Heaven in a Rage*, 234–35; Gerald Carson, *Men, Beasts and Gods* (New York: Scribner's, 1972), 103; Eric Shelman and Stephen Lazoritz, *The Mary Ellen Wilson Child Abuse Case and the Beginning of Children's Rights in 19th Century America* (Jefferson, NC: McFarland, 2005).
8. See the reference to Aquinas in chapter 5, above, pp. 219–222.
9. See Farley Mowat, *Never Cry Wolf* (Boston: Atlantic Monthly Press, 1963), and Lorenz, *King Solomon's Ring*, 186–89. I owe the first reference to Mary Midgley, "The Concept of Beastliness: Philosophy, Ethics and Animal Behavior," *Philosophy* 48 (184): 111–35 (1973).
10. See, in addition to the references above, works by Niko Tinbergen, Jane Goodall, George Schaller, and Irenaus Eibl-Eibesfeldt.
11. Brock Bastian, et al., "Don't mind meat? The denial of mind to animals used for human consumption," *Personality and Social Psychology Bulletin* 38 (2): 247–56 (2012).
12. See the references to Paley and Franklin in chapter 5, above, 233–234.
13. The phrase is from Tennyson's poem *In Memoriam A.H.*, written in 1850.
14. For the question raised about alleviating animal suffering, see, for example, D. G. Ritchie, *Natural Rights* (London: Allen & Unwin, 1894), reprinted in Regan and Singer, eds., *Animal Rights and Human Obligations*, 183.
15. Aldo Leopold, "Thinking Like a Mountain," *Sand County Almanac* (1949; reprint, New York: Oxford University Press, 2020.
16. The term was first used by Yew-Kwang Ng, "Towards welfare biology: Evolutionary economics of animal consciousness and suffering," *Biology and Philosophy* 10 (3): 255–85 (1995). See also Catia Faria and Oscar Horta, "Welfare Biology," in Bob Fischer, ed., *The Routledge Handbook of Animal Welfare* (New York: Routledge, 2019), 455–66; and Asher Sorel, et al., "The Case for Welfare Biology," *Journal of Agricultural and Environmental Ethics* 34:7 (2021), https://doi.org/10.1007/s10806–021–09855–2.
17. Catia Faria and Oscar Horta, "Welfare Biology"; see also Catia Faria, *Animal Ethics in the Wild* (Cambridge: Cambridge University Press, 2022).
18. For these suggestions I am indebted to Yip Fai Tse, Oscar Horta, Catia Faria and Wild Animal Initiative.
19. See Catia Faria, *Animal Ethics in the Wild*, p. 84, and the sources cited there.

20. See Oscar Horta, "Animal suffering in nature: The case for intervention," *Environmental Ethics* 39 (3): 261–79 (Fall 2017); and Catia Faria, *Animal Ethics in the Wild.*
21. Jane Capozelli, et al., "What is the value of wild animal welfare for restoration ecology?," *Restoration Ecology* 28: 267–70 (2020).
22. Brigid Brophy, "In Pursuit of a Fantasy," in Godlovitch and Harris, eds., *Animals, Men and Morals*, 132.
23. See Cleveland Amory, *Man Kind?: Our Incredible War on Wildlife* (New York: Harper and Row, 1974), 237.
24. Lewis Gompertz, *Moral Inquiries on the Situation of Man and of Brutes* (London, 1824, reprint, Lewiston, NY: Edwin Mellen Press, 1997).
25. For a powerful account of the cruelties inherent in the Australian wool industry, see Christine Townend, *Pulling the Wool* (Sydney: Hale & Iremonger, 1985). Although this book is nearly forty years old, some of the worst practices, such as mulesing without anesthetic, are still legal and widely practiced. See Wool With a Butt, "Our Progress: Timeline to End Mulesing," April 12, 2022: woolwithabutt.four-paws.org/issues-and-solutions /timeline-to-end-mulesing.
26. The view of plants as actively responding to their environment goes back to Darwin, but it was introduced to many contemporary readers by Peter Wohlleben's bestseller *The Hidden Life of Trees*, (Vancouver, Greystone, 2016). There is now an extensive literature in this area. For a review, see Sergio Miguel-Tomé, and Rodolfo Llinás, "Broadening the definition of a nervous system to better understand the evolution of plants and animals," *Plant Signaling & Behavior* 16 (10): article 1927562(2021).
27. I owe the parallel with artificial intelligence invoked in this paragraph to Stevan Harnad. See his blog Skywritings, July 14, 2022, generic.wordpress .soton.ac.uk/skywritings/category/sentience/.
28. Richard Wasserstrom, "Rights, Human Rights and Racial Discrimination," in A. I. Melden, ed., *Human Rights* (Belmont, CA: Thomson Wadsworth, 1970), 106.
29. W. Frankena, in Richard Brandt, ed., *Social Justice* (Englewood Cliffs, NJ: Prentice-Hall, 1962), 23; H. A. Bedau, "Egalitarianism and the Idea of Equality," in J. R. Pennock and J. W. Chapman, eds., *Nomos IX: Equality*, (New York, Atherton Press, 1967); G. Vlastos, "Justice and Equality," in *Social Justice*, 48.
30. John Rawls, *A Theory of Justice* (Cambridge, MA: Harvard University Press/ Belknap Press, 1972), 510. For another example, see Bernard Williams, "The Idea of Equality," in P. Laslett and W. Runciman, eds., *Philosophy, Politics and Society*, second series (Oxford, Blackwell, 1962), 118.
31. Bernard Williams, "The Human Prejudice," in Bernard Williams, *Philosophy as a Humanistic Discipline*, A. W. Moore, ed. (Princeton University Press, 2006), 152.

32. For an example, see Stanley Benn's "Egalitarianism and Equal Consideration of Interests," in J. R. Pennock and J. W. Chapman, eds., *Nomos IX: Equality*, (New York, Atherton Press, 1967), 62ff.
33. Shelly Kagan, *How to Count Animals, more or less* (Oxford University Press, 2019); see also Shelly Kagan, "What's wrong with speciesism?" *Journal of Applied Philosophy* 33: 1–21 (2016).
34. Eric Schwitzgebel, Bradford Cokelet, and Peter Singer, "Do ethics classes influence student behavior? Case study: Teaching the ethics of eating meat," *Cognition* 203: article 104397 (October 2020); Eric Schwitzgebel, Bradford Cokelet, and Peter Singer, "Students Eat Less Meat After Studying Meat Ethics," *Review of Philosophy and Psychology* 1–26 (November 2021).
35. Philipp Schönegger and Johannes Wagner, "The moral behavior of ethics professors: A replication-extension in German-speaking countries," *Philosophical Psychology* 32(4): 532–59 (2019); see also Andrew Sneddon, "Why do ethicists eat their greens?", *Philosophical Psychology* 33: 902–923 (2020).
36. Efforts to find this quote in Gandhi's writings have failed. The earliest close match found is from Nicholas Klein at a convention of the Amalgamated Clothing Workers of America in 1918. For details of this and other related comments, including one from Gandhi, see Quote Investigator: quoteinvestigator.com/2017/08/13/stages/.
37. For an account of Spira's successes, see Peter Singer, *Ethics into Action* (Lanham, MD: Rowman and Littlefield, 2019).
38. www.hsi.org/news-media/fur-trade/; www.furfreealliance.com/republic-of-ireland-bans-fur-farming.
39. Vanessa Friedman, "The California Fur Ban and What it Means for You," *New York Times*, October 14, 2019.
40. Humane Society International, "Inhumane rodent glue traps to be banned in England following unanimous vote in House of Lords," www.hsi.org/news-media/inhumane-rodent-glue-traps-to-be-banned-in-england; www.peta.org/features/join-campaign-glue-traps.
41. See Paola Cavalieri and Peter Singer, eds., *The Great Ape Project: Equality beyond humanity* (London: Fourth Estate, 1993).
42. For further information on this and other cases, see Macarena Montes Franceschini, "Animal Personhood: The Quest for Recognition," *Animal and Natural Resource Law Review* XVII: 93–150 (July 2021).
43. Animal Equality, "European Parliament Votes to Ban Cages on Farms," June 10, 2021, animalequality.org/news/european-parliament-votes-ban-cages/.
44. Animal Welfare Institute, "Legal Protections for Animals on Farms," May 2022, awionline.org/sites/default/files/uploads/documents/22-Legal-Protections-Farm.pdf.
45. Compassion in World Farming, "All of Top 25 U.S. Food Retailers Go Cage-Free," July 18, 2016, www.ciwf.com/blog/2016/07/all-of-top-25-us-food-retailers-go-cage-free.

46. Beth Kowitt, "Inside McDonald's Bold Decision to Go Cage-Free," *Fortune* August 18, 2016.
47. Natalie Berkhout, "'Largest and Longest' Global Cage-Free Campaign a Success," *Poultry World* 22 (September 2021).
48. USDA AMS Livestock & Poultry Program, Livestock, Poultry, and Grain Market News Division, *Egg Markets Overview* July 29, 2022.
49. "Plant-based Foods Market to Hit $162 Billion in Next Decade, Projects Bloomberg Intelligence, "August 11, 2021, www.bloomberg.com/company/press/plant-based-foods-market-to-hit-162-billion-in-next-decade-projects-bloomberg-intelligence/.
50. Winston Churchill, "Fifty Years Hence," *Strand Magazine* December 1931, www.nationalchurchillmuseum.org/fifty-years-hence.html.
51. Good Food Institute, *State of the Industry Report: Cultivated Meat and Seafood,* 2022, gfi.org/resource/cultivated-meat-eggs-and-dairy-state-of-the-industry-report/.

INDEX

ABOUT THE AUTHOR

Peter Singer was born in Melbourne, Australia, in 1946, and educated at the University of Melbourne and the University of Oxford. After teaching in England, the United States, and Australia, he has, since 1999, been Ira W. DeCamp Professor of Bioethics in the University Center for Human Values at Princeton University. Singer first became well-known internationally after the publication of *Animal Liberation* in 1975. His other books include: *Practical Ethics*, *The Expanding Circle*, *The Ethics of What We Eat* (with Jim Mason), *The Point of View of the Universe* (with Katarzyna de Lazari-Radek), *The Most Good You Can Do*, *Ethics in the Real World*, *Utilitarianism: A Very Short Introduction* (with Katarzyna de Lazari-Radek), and *Why Vegan?* The publication of *The Life You Can Save* in 2009 led him to found a nonprofit organization of the same name that recommends the most effective charities assisting people in extreme poverty. In 2012 Singer was made a Companion of the Order of Australia, the country's highest civic honor, and in 2021 he was awarded the $1 million Berggruen Prize for Philosophy and Culture, which he divided between nonprofit organizations working for animals and those seeking to reduce global poverty.